Dynamic Analysis of the Urban Economy

STUDIES IN URBAN ECONOMICS

Under the Editorship of

Edwin S. Mills
Princeton University

Dynamic Analysis
of the Urban Economy

TAKAHIRO MIYAO

Department of Economics
University of Southern California
Los Angeles, California

 1981

ACADEMIC PRESS
A Subsidiary of Harcourt Brace Jovanovich, Publishers
New York   London   Toronto   Sydney   San Francisco

ACADEMIC PRESS, INC.
111 Fifth Avenue, New York, New York 10003

United Kingdom Edition published by
ACADEMIC PRESS, INC. (LONDON) LTD.
24/28 Oval Road, London NW1 7DX

Library of Congress Cataloging in Publication Data

Miyao, Takahiro.
    Dynamic analysis of the urban economy.

    (Studies in urban economics)
    Includes bibliographies and index.
    1. Cities and towns--Growth. 2. Urban economics.
3. Externalities (Economics) I. Title. II. Series.
HT321.M58   330.9173'2   80-70590
ISBN 0-12-501150-4   AACR2

PRINTED IN THE UNITED STATES OF AMERICA

81 82 83 84   9 8 7 6 5 4 3 2 1

To my parents

Contents

PART 4   CONGESTION AND AGGLOMERATION

Foreword

One of the pleasures of being an economic theorist is the discovery that some new aspect of social life, superficially quite different from the explicit haggling of the marketplace, will actually yield to the methods of economic theory. I can remember enjoying that feeling when I first began to think about the geography of economic activity inside a city, and its codetermination along with the pattern of rents, transportation flows, and congestion. You do not have to be a pioneer to have this experience, although it helps to be a little ignorant of the literature. In this case, the tradition begins with von Thünen, and its most important contemporary representatives are Alonso, Beckmann, Mills, Mohring, Muth, Vickrey, and now Miyao.

Modern theoretical urban economics has been almost entirely a study of equilibrium configurations – patterns of residential and/or industrial location with the property that no one wishes to move. Professor Miyao has now added some explicit dynamics. This takes two forms. In the early chapters, and occasionally elsewhere, he is concerned with adjustment dynamics, the response of the model to displacements from the equilibrium pattern. Under some assumptions, he finds that the equilibrium is stable, and he is able to use stability conditions to do comparative statics in the manner of the Samuelson Correspondence Principle. In an interesting passage, he shows that other assumptions, especially those embodying externalities such as intergroup aversion or intragroup reinforcement, can result in a kind of instability. This is a lot like what is called neighborhood "tipping," and it is a merit of Miyao's analysis that it focuses attention on the parameters that govern the intensity of this instability.

In later chapters, Professor Miyao turns to a kind of equilibrium dynamics, rather like growth theory, and studies how an equilibrium city might evolve in response to exogenous forces, of which population growth is the most obvious example. Here, too, he finds a number of interesting results, including a sort of "golden rule" for transportation investment—perhaps not surprising in view of the analogy to growth theory.

I suppose the history of economic thought teaches us that where there is equilibrium, there is dynamics. What I find remarkable about the step that Miyao has taken is that he has had the imagination, ingenuity, and energy actually to find the dynamics in what must at first have seemed like an impenetrable thicket. Similarly, I remember being astonished, back in Professor Miyao's graduate-student days, that anyone should have the courage to attack a two-dimensional version of a problem that Vickrey and I had analyzed in one dimension. For the end result, a victory of clarity of mind, see Chapters 12 and 13 of this book.

When a complicated aspect of social life does yield to the methods of economic theory, a good theorist realizes that it rarely yields everything. The typical experience is that the first-order effects have the right sign, and the orders of magnitude accord with common observation. I think that such is the case with the "new urban economics" exemplified in this book. That is encouraging when it happens—certainly more encouraging than the opposite outcome!—and it gives one some confidence that the theoretical model may be useful in singling out the important effects and key parameters that govern the broad outlines of real phenomena.

Of course, the creative effort of serious application still remains. But even for the theorist there will—one hopes—always be interesting effects of more delicate order to be smuggled into the model and analyzed. Every reader of this book will have favorite examples. Someone who has recently moved from suburb to city, and likes much but not all of what he sees, might be interested in modeling "gentrification." Someone who has recently visited Memphis might be interested in modeling a city built on a river wide enough to be an effective residential barrier. I happen to fall into both categories. Professor Miyao might some day look into the statics and dynamics of transportation improvements that are highly localized and could thus possibly serve as a planning device. Good cookbooks encourage people to try their own recipes. So, as Julia Child says, "Bon appetit."

ROBERT M. SOLOW

Preface

This volume is intended to provide a dynamic analysis of economic activities in urban areas. The main theme of the book is dynamics, although related subjects such as existence, uniqueness, comparative statics, and optimality are also discussed. I believe that a dynamic analysis is urgently needed in urban economics because, despite growing interest in understanding the dynamism of urban activities, particularly from public policy viewpoints, most of the existing literature on urban economics has dealt only with determination and characterization of spatial equilibrium in static urban systems and thus has failed to shed light on the dynamic aspect of economic activities in and around urban areas. It might be said that, except for a few journal articles on urban growth and housing, the whole field of urban economics is still awaiting a systematic application of the dynamic method, which has been fully developed and widely used in economics.

In this book, I attempt to take a step toward a systematic treatment of the dynamic aspect of business and residential activities in the urban economy. The main purposes of this study are (1) to obtain some insight into the dynamic processes of complex urban relationships by constructing and analyzing simple dynamic models of the urban economy and (2) to contribute to the development of what might be called "dynamic urban economics" within the framework of general dynamic economics, as laid out by Samuelson, Solow, and others, and thereby provide a sound theoretical basis for understanding and predicting the dynamic processes of the urban economy. As it turns out, urban economics is a very fertile field for application of the conventional dynamic analysis that has been developed in connection with general equilibrium theory and economic growth theory. I hope that this book has reaped some good crops from the field and will serve as an example to show the richness of the area of dynamic urban economics.

The Introduction is a preview of my basic ideas about dynamics. Theoretical analyses of dynamic urban systems are given in the subsequent chapters. In Part 1 account is taken of the dynamic stability property of spatial equilibrium and its relation to comparative statics. The effects of various kinds of externalities on the dynamic property of the urban economy are discussed in Part 2. The long-run growth processes of the urban economy and their optimality property are presented in Part 3. In Part 4, the optimal size and configurations of an urban area in connection with agglomeration economies and traffic congestion are given.

This book grows out of my dissertation—submitted to the Massachusetts Institute of Technology in 1974—and my articles published in various journals. It is addressed primarily to colleagues in the field of urban economics and in other fields of economics. The book may also be used as supplementary reading in graduate urban economics courses. A reader who has some knowledge of calculus and linear algebra should be able to follow the main argument of this book. Familiarity with elementary set theory may be useful but is not required. References are given at the end of each chapter.

Acknowledgments

Throughout the course of my research, I have received helpful guidance and encouragement from my former thesis supervisor, Robert M. Solow, to whom I should like to express my deep appreciation. It also gives me great pleasure to acknowledge my intellectual debt to Paul A. Samuelson and Franklin M. Fisher, whose influence can be easily detected in this book. Thanks are due to Richard Arnott, Jan Brueckner, David Knapp, Edwin Mills, Leon Moses, Harry Richardson, Jerome Rothenberg, Perry Shapiro, and William Wheaton, who have given me comments and suggestions along the way. I must also acknowledge a debt of gratitude to my former teachers at Keio University, Yoshindo Chigusa and Masao Fukuoka, and to my friends, Koichi Hamada and Hajime Oniki.

Finally, special thanks are reserved for my wife, Mariko, whose cooperation and understanding made it possible for me to write this book.

Introduction

Cities are born to be dynamic. They are growing, changing, decaying, and redeveloping. Most people prefer urban life for its dynamic feelings, activities, and opportunities, which are not present in rural life. A city will lose its meaning and attractiveness when it becomes stagnant and motionless. In a sense, cities define themselves by their dynamism. In economic terms, one might define a city as a "settlement that consistently generates its economic growth from its own local economy (see Jacobs, 1970, p. 262)." There is no doubt that dynamism is one of the most important characteristics of urban processes and that no study of the urban economy could be complete without taking account of its dynamics and growth.

In this book we attempt to offer a dynamic analysis of the urban economy. The present work may be regarded as a straightforward application of well-established dynamic methods to urban economics for the purpose of analyzing the dynamic processes of urban economic activities. By using the conventional dynamic approach, this study is intended to contribute to the development of "dynamic urban economics" within the context of general dynamic economics in order to provide a sound theoretical basis for understanding the working of the urban economy.

As is well known, dynamic economics is meant to include two different, although closely related, types of analyses: the analysis of the dynamic stability property of equilibrium in a static economy and the analysis of the dynamic processes of a growing economy. The former might be called the "stability analysis" and the latter the "growth analysis." The stability analysis, developed in connection with general equilibrium analysis, assumes

1

some adjustment mechanism to determine the dynamic paths of price or quantity variables toward an equilibrium in an economic system with given endowments of resources and technology, whereas the growth analysis offers the equation of dynamic motion for an economic system with its endowments of resources and/or technology changing over time. In this book, we are concerned with the both types of analyses as applied to urban economics.

It may be useful to illustrate how we apply the stability and growth analyses to urban location theory by using simple diagrams. First, our basic idea about application of the stability analysis can be seen in Fig. 1, where two "bid rent" curves are drawn in a monocentric city with its center at the origin. [For the concept of bid rent, see Alonso (1964).] The two bid rent curves represent the maximum rents which two groups of firms (or residents) can pay to landowners at each distance from the city center. As is well known, a group, say group 1, demanding relatively less land relative to marginal transport cost will have a steeper bid rent curve (AEB) than that (CED) of a group, say group 2, with relatively more demand for land relative to marginal transport cost. Since land should go to the highest bidder, at least in equilibrium, group 1 will occupy the inner segment of the city up to the distance x^* from the center, whereas group 2 will be accommodated in the outer segment of the city. As a result, the equilibrium market rent

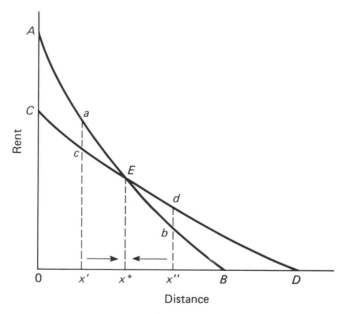

FIGURE 1

curve can be obtained by taking the upper envelope of the two bid rent curves, i.e., the curve AED.

Now, consider a disequilibrium situation with the initial position of the boundary between the two groups given arbitrarily, say at x', and the market rent curve represented by $AacED$. This means that group 1 is actually located between 0 and x', while group 2 is housed outside x', although between x' and x^* group 1 would be willing to pay higher rents than group 2 which actually occupies that segment of the city—a disequilibrium situation. Then it seems natural to suppose that some adjustment will take place so as to move the actual boundary between the two groups toward its equilibrium position x^*, as some members of group 1 take over the land area between x' and x^* gradually from group 2. For analytical convenience, we assume this takeover to occur continuously from x' to x^*.

If, on the other hand, the initial position of the boundary is given at x'', the market rent curve is represented by $AEbdD$, and the segment between x^* and x'' is occupied by group 1 and not by the highest bidders, i.e., group 2, for that segment of land. Then the actual position of the boundary tends to move in, continuously toward the equilibrium position, as some members of group 2 move into that segment gradually from x'' to x^*.

Thus, with the two bid rent curves as given in Fig. 1, the equilibrium boundary position seems to be dynamically stable, according to the dynamic adjustment process specified above. Our explanation so far, however, has been rather intuitive and based on partial equilibrium analysis, in that the position of the bid rent curves are fixed regardless of changes in the boundary position, which should actually affect the positions of the bid rent curves. It remains to be seen if dynamic stability can be established in a general equilibrium model which takes account of all direct and indirect effects of boundary changes during the course of dynamic adjustment. In subsequent chapters, we shall rigorously prove the dynamic stability property of spatial equilibrium in general equilibrium models of urban location with many groups of firms and residents.

Let us turn to the growth analysis as applied to urban location theory. Here again, we use a simple diagram to illustrate the main point of our argument. In Fig. 2, there is a monocentric city which is occupied by only one group of firms (or residents) and no alternative use of land is assumed for the sake of simplicity. At time 0, the group's bid rent curve, which is also the market rent curve, is given by the curve AB and the outer boundary of the city is located at x^0, where the opportunity cost of land is assumed to be zero. As time passes, the city may be changing in terms of its population, transport cost, demand for its output, etc. As a result, the market rent curve may shift inward or outward, depending on the direction of changes in those factors.

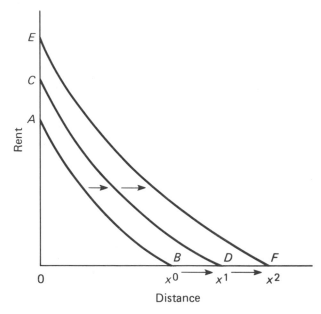

FIGURE 2

Specifically, an increase in total population coupled with a decrease in unit transport cost, say, due to social investment in urban transportation, will lead to an outward shift in the position of the market rent curve, as depicted in Fig. 2. This occurs because the reduced cost of transportation will enable existing firms (residents) to pay higher rents for each location and competition with new firms (new residents) will force them to do so. Thus, in Fig. 2 the market rent curve shifts outward from AB at time 0 to CD at time 1, and then EF at time 2, and correspondingly the city expands in size from x^0 to x^1, and then to x^2.

It is natural to ask whether the urban growth process as described can continue for a reasonably long period of time. As it turns out, we are able to construct simple models of urban growth, in which the total land area and the total population of the city along with other factors, if any, are growing steadily at the same rate, and such a steady-state balanced-growth equilibrium can be sustained indefinitely, given the availability of open space for the steady expansion of the urban area. In the following chapters, we shall examine the existence, uniqueness, and stability properties of a balanced growth equilibrium and explore the optimality property of such a growth path by applying the ordinary method used in economic growth theory.

Having explained our basic ideas about dynamic analysis, we are now in a position to give an overview of the main contents of the book. There are four parts on theoretical studies of urban systems with special emphasis on dynamics, growth, and agglomeration.

In Part 1, we investigate the dynamic stability property of spatial equilibrium and its relation to comparative statics in urban location models. In Chapter 1, the dynamics of industrial location is studied by setting up a production location model of the von Thünen type with many industries and introducing dynamic adjustment processes of wages and industrial boundaries. Local stability of spatial equilibrium is proved and some comparative static results are obtained. An application of the LeChatelier principle is also explored. In Chapter 2, we derive some dynamic and comparative static results in a residential location model of a closed city with many classes of households, where the city is closed in the sense that the number of households in each class is fixed exogenously, while their utility level is determined endogenously. In Chapter 3, an urban location model with both production and residential activities is presented in order to take account of the interaction of industrial and residential locations. A simple model with one industry and one household class is used to illustrate how their interaction complicates conditions for dynamic stability of spatial equilibrium. In Chapter 4, dynamics and comparative statics are considered regarding an open city in which the level of utility is exogenous while the size of each household class is endogenous, in contrast to the closed city model in Chapter 2.

In Part 2, various kinds of externalities are introduced into urban location models, and we examine how the presence of externalities affects the dynamic property of the urban economy. Our results point to the fact that certain types of externalities will weaken, if not totally destroy, the stable nature of the urban economy. Neighborhood externalities among two groups of residents are considered in Chapter 5, and we prove that an interior equilibrium with two groups coexisting in the city may be stable or unstable, according as the degree of externality is relatively small or large. In Chapter 6, we develop a general model of probabilistic location choice by many types of residents who interact among themselves in the presence of neighborhood effects. The existence, uniqueness, and stability of an equilibrium are rigorously proved. The equilibrium is shown to be stable or unstable, depending on whether the degree of externality is relatively small or large, just as in Chapter 5. In Chapter 7, we focus on production externalities such as air pollution in a simple model of an open city with both production and residential activities. If utility functions and production functions are Cobb–Douglas, an equilibrium exists, but its uniqueness and stability may or may not follow, depending on the degree of externality, as in the previous chapters.

In Part 3, urban growth models which incorporate both spatial elements and growth factors explicitly are developed to study the long-run growth processes of the urban economy. In Chapter 8, we present a growth model of an industrial city with population growth and transportation improvements. The existence, uniqueness, and global stability of a balanced growth equilibrium are proved. Also, a golden rule result is found with regard to optimal taxation for transportation improvements. In Chapter 9, residential growth is considered in a residential location model, and the existence, uniqueness, and global stability of a balanced growth equilibrium are proved. A golden rule result is also obtained. In Chapter 10, we set up a demand-oriented model of urban growth with unemployment, in contrast to the growth models with full employment assumed in the previous chapters. Dynamic and comparative static analyses are carried out in order to investigate the long-run behavior of such variables as unemployment, income, and population in the urban economy in response to exogenous demand growth for output. It is shown that the long-run property of the model depends crucially on the sensitivity of the rate of nominal wage increase to changes in the rate of price increase in the urban economy. In Chapter 11, we introduce agglomeration economies into a two-sector model of rural–urban migration of the Harris–Todaro type with urban unemployment. It is proved that with agglomeration economies in the urban sector, there are likely to exist multiple steady-state equilibria, some of which are unstable in the long run.

In Part 4, we investigate optimal land use and the optimal degree of concentration in two-dimensional noncircular cities with traffic congestion and agglomeration economies. A rectangular city with a grid road system where alternative route choice is open to individual road users is offered in Chapter 12, and we show that the conventional cost–benefit analysis based on market rents leads to excessive road-building especially near the city boundaries, rather than near the city center. It is also proved that free entry of competitive firms into the city tends to yield an excessively large city size. In Chapter 13, the problem of optimal concentration of business activities is studied by setting up a model of a multicentric city with the number of city centers treated as a variable. An optimal number of city centers is determined by the relative forces of agglomeration economies and traffic congestion. Some numerical examples are provided to show that with a modest degree of agglomeration economies the optimal number of city centers will gradually increase with the volume of commuting traffic relative to the volume of business traffic.

REFERENCES

Alonso, W. (1964). *Location and Land Use*. Harvard Univ. Press, Cambridge, Massachusetts.
Jacobs, J. (1970). *The Economy of Cities*. Random House, New York.

PART 1

STABILITY OF SPATIAL EQUILIBRIUM

CHAPTER 1

Industrial Location

In this chapter we examine the dynamic property of spatial equilibrium in a general equilibrium model of production.[1] The model includes many industries located in a circular city with a central market place to which all products must be transported. This type of model, being analogous to von Thünen's model of a monocentric region, has recently been adopted and widely used in urban economics (e.g., see Alonso, 1964; Beckmann, 1972; Mills, 1967).

First, we derive conditions for spatial equilibrium under perfect competition, and introduce dynamic adjustment processes in disequilibrium situations. Specifically, we consider two alternative adjustment processes: a gradual adjustment of boundaries between industrial zones in the land market, and a gradual adjustment of the wage rate in the labor market. In both cases the equilibrium is shown to be dynamically stable under the condition that at each boundary between two industrial zones the industry located in the inner zone should have a higher ratio of marginal transport cost to land per unit of output than the industry located in the outer zone. In fact, this stability condition is the same as von Thünen's condition for locational equilibrium.

Then some comparative static results are derived from our dynamic analysis by applying the Correspondence Principle in the present context.[2] It is proved that an increase in the total amount of labor available in the

---

[1] This chapter is largely based on Miyao (1977, 1980).

[2] For a general treatment of the Correspondence Principle, see Samuelson (1947).

city will lead to (1) a decrease in the wage rate, (2) an outward *or* inward movement of a boundary between two industrial zones inside the city, according as the industry located in the inner zone has a higher *or* lower labor–land ratio than the industry located in the outer zone at the boundary, and (3) an outward movement of the outer (exterior) boundary of the city. Exactly the opposite effect results from the availability of more land for production inside the city. It is also shown that an increase in the opportunity cost of land will give rise to (1) a fall in the wage rate, (2) an outward *or* inward movement of a boundary between two industrial zones inside the city, according as the inner industry has a higher *or* lower labor–land ratio than the outer industry at the boundary, and (3) an inward movement of the outer boundary of the city.

Furthermore, regarding the effect of a change in the opportunity cost of land, we ask how the rate of change of the outer boundary of the city will be affected when some or all of the boundaries between industries inside the city are fixed at their original positions. This question can be answered by applying the LeChatelier Principle in the present context.[3] That is, the effect of a given change in the opportunity cost of land on the position of the outer boundary of the city will be greater, as more boundaries between industries inside the city are allowed to vary in response to the change in the opportunity cost of land.

1.1 The Production Model

Consider a city characterized by von Thünen's homogeneous plane with a single central market to which producers must transport all their products from their production points in the city. For each producer, unit transport cost, i.e., the cost of transporting one unit of output, depends only on distance and is an increasing function of distance from the central market. There are assumed to be a finite number of industries producing different kinds of commodities subject to constant-returns-to-scale production functions of the ordinary neoclassical type, using land and labor for production. All producers in the same industry, producing the same kind of commodity, have identical production functions and identical transport cost functions, whereas those functions may be different from industry to industry. We also suppose that the city considered here is a small open economy, freely trading

[3] See Samuelson (1947, 1960) on the LeChatelier Principle.

with the national or international economy, so that all product prices prevailing at the central market are given exogenously.[4]

In equilibrium, industries are assumed to locate themselves in concentric rings, or "zones," around the central market in such a way that each zone is filled exclusively with producers in a single industry, while one industry is allowed to occupy two zones or more separated by some other industries. Without loss of generality, however, we can set the number of zones equal to the number of industries by redefining an "industry" as a group of producers producing the same kind of commodity *and* occupying the same zone in equilibrium. Let both industries and zones be numbered in ascending order of distance from the central market so that the ith zone from the center is called zone i and occupied by industry i for $i = 1, \ldots, m$, where m is the total number of industries in the city.

Because of the assumptions that only land and labor are used for production and that constant returns to scale prevail, we may regard the *minimum* cost of producing one unit of product by industry i as a function of the rental price of land r and the wage rate w,

$$C_i - C_i(r, w) \qquad (i = 1, \ldots, m),$$

which is called industry i's unit cost function. In equilibrium under perfect competition, all producers should make zero profit at every location so that industry i's unit production cost C_i equals the net price of its product, i.e., product price p_i minus unit transport cost q_i,

$$C_i[r_i(x), w] = p_i - q_i(x) \qquad (i = 1, \ldots, m), \tag{1}$$

where unit transport cost $q_i(x)$ is a strictly increasing function of distance x, and p_i and w are independent of x. In effect, Eq. (1) defines industry i's "bid rent" $r_i(x)$ at each distance x, which is the *maximum* rent that producers in industry i are willing to pay for each location without incurring losses. In equilibrium, the fact that industry i occupies zone i exclusively requires that industry i's bid rent be among the highest, i.e., at least equal to the bid rent of any other industry, everywhere in zone i:

$$r_i(x) \geqq r_j(x) \qquad \text{for all} \quad j = 1, \ldots, m, \quad \text{for} \quad x_{i-1} \leqq x \leqq x_i,$$

where x_{i-1} and x_i are the inner radius and the outer radius of zone i, respectively. This means that the segment of the market rent function between x_{i-1} and x_i is represented by industry i's bid rent function in zone i for $i = 1, \ldots, m$.

---

[4] These assumptions are essentially the same as those of Solow (1973). In fact, as far as its static aspect is concerned, the present model is an extension of the Solow model, which involves only one industry.

In order to know more about bid rents, we note that according to the well-known properties of the unit cost function, the partial derivatives of C_i with respect to r_i and w are equal to industry i's demand for land and labor, respectively, per unit of output.[5] Assuming that these partials to be always positive, i.e.,

$$C_{ir} \equiv \partial C_i/\partial r_i > 0, \qquad C_{iw} \equiv \partial C_i/\partial w > 0,$$

we can apply the implicit function theorem to Eq. (1) and obtain industry i's bid rent r_i at x as a function of $p_i - q_i(x)$ and w, i.e., $r_i(x) = R_i[p_i - q_i(x), w]$, or in short

$$r_i(x) = r_i(x, w) \qquad \text{for} \quad x_{i-1} \le x \le x_i \qquad (i = 1, \ldots, m), \tag{2}$$

with the partial derivatives

$$r_{ix}(x) \equiv \partial r_i(x)/\partial x = -q_i'(x)/C_{ir}(x) = -q_i'(x)/h_i(x) < 0, \tag{3}$$

$$r_{iw}(x) \equiv \partial r_i(x)/\partial w = -C_{iw}(x)/C_{ir}(x) = -n_i(x) < 0, \tag{4}$$

where h_i is industry i's demand for land per unit of output, and n_i is industry i's demand for labor per unit of land. Equation (3) confirms the well-known fact that the market rent gradient is negative and its absolute value is equal to the ratio of marginal transport cost to land. A city with a typical shape of bid rent functions is illustrated in Fig. 1.1.

Next, we turn to the labor market and derive a condition for full employment. Suppose that at each distance x a fraction, say g, of the total area of a thin ring with inner radius x and outer radius $x + dx$ is available for production, and g is a known function of x, where $g(x) > 0$ for all x. For example, in Fig. 1.1 we have $g(x) = \frac{1}{2}$ for all x. Then industry i's demand for labor in the thin ring at x is equal to $n_i(x)g(x)2\pi x\, dx$. Here we note that n_i is defined in Eq. (4) as industry i's labor–land ratio and is a function of r_i and w,

$$n_i(x) = C_{iw}/C_{ir} = n_i[r_i(x), w] \qquad \text{for} \quad x_{i-1} \le x \le x_i \quad (i = 1, \ldots, m). \tag{5}$$

If the total amount of labor available in the city is a given positive constant, say N, the full employment condition for the city as a whole can be written as

$$2\pi \sum_{i=1}^{m} \int_{x_{i-1}}^{x_i} n_i(x)g(x)x\, dx = N, \tag{6}$$

where the radius of the central (disk) market x_0 is given exogenously as a nonnegative constant. In Fig. 1.1, the market reduces to a point so that $x_0 = 0$.

---

[5] For the concept and properties of the cost function, refer to Diewert (1974), Shephard (1953, 1970), and Varian (1978).

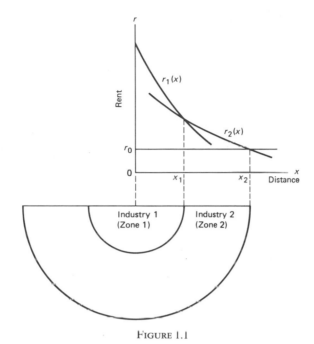

FIGURE 1.1

Finally, to attain equilibrium in the land market we need what might be called the "market rent condition" requiring the market rent function to be continuous everywhere in the city, particularly at every boundary between industrial zones x_i ($l = 1, \ldots, m - 1$) and the outer boundary of the city x_m. It should be clear from Fig. 1.1 that in equilibrium industry i's bid rent function in zone i represents the segment of the market rent function between x_{i-1} and x_i. The market rent condition is thus

$$r_i(x_i) = r_{i+1}(x_i), \qquad r_m(x_m) = r_0 \qquad (i = 1, \ldots, m - 1), \tag{7}$$

where r_0 is the opportunity cost of land, e.g., agricultural land rent, which is assumed to be given exogenously.

Our model consists of Eqs. (1) and (5)–(7), which together can determine the equilibrium functional forms of $r_i(x)$ and $n_i(x)$ ($i = 1, \ldots, m$) and the equilibrium values of w and x_i ($i = 1, \ldots, m$) under some assumptions. Since we are interested primarily in dynamics and comparative statics, we simply assume the existence of an equilibrium as follows. *There are assumed to exist a set of functions, $r_i^*(x)$ and $n_i^*(x)$ ($i = 1, \ldots, m$), and a set of values w^* and x_i^* ($i = 1, \ldots, m$) such that Eqs. (1) and (5)–(7) are simultaneously satisfied.* The asterisk may be omitted when there is no possible confusion.

1.2 Boundary Adjustment

Let us analyze the dynamic stability property of the equilibrium in our model. We examine how the boundaries between industries and the wage rate will move through time in response to a small displacement from equilibrium by assuming some adjustment processes in disequilibrium. Specifically, we introduce the following two alternative processes of dynamic adjustment: (1) a gradual adjustment of the boundaries responding to rent differences in the land market, and (2) a gradual adjustment of the wage rate responding to the excess demand for labor in the labor market, where all the other variables are assumed to be fully adjusted at each point in time. The former process is analyzed in this section, while the latter is considered in the next section.

Suppose initially there are arbitrary, but sufficiently small, changes in some or all of the boundaries from their equilibrium positions and, as a result, the market rent condition (7) is no longer satisfied; i.e., the overall rent function for the city as a whole is no longer continuous at initially given boundary positions.[6] Here we postulate that in the presence of some frictional elements in the land market, each boundary will be adjusted gradually through time in such a way that the position of a boundary moves outward *or* inward, according as the land rent just inside the boundary is higher *or* lower than that just outside the boundary.[7] It is also natural to assume that the rate of boundary adjustment with respect to time will decline as the rent difference at the boundary becomes smaller. On the other hand, all the other variables r_i, n_i, and w are assumed to be fully adjusted so as to satisfy Eqs. (1), (5), and (6) instantaneously for each given set of values of x_i. Then, the dynamic process of boundary adjustment can be expressed as

$$\begin{aligned}
\dot{x}_i &= f_i[r_i(x_i) - r_{i+1}(x_i)] \qquad (i = 1,\ldots,m-1), \\
\dot{x}_m &= f_m[r_m(x_m) - r_0],
\end{aligned} \tag{8}$$

where the dot denotes differentiation with respect to time, and

$$f_i'(\ ) > 0, \qquad f_i(0) = 0 \qquad (i = 1,\ldots,m).$$

In order to examine the dynamic property of system (8), we must know exactly how bid rents depend on boundary positions. Let us define the expres-

---

[6] It should be noted that we are dealing with disequilibrium to such a small degree that the ordering of industries will not be affected.

[7] See the Introduction for an intuitive explanation of this adjustment process.

sion on the left-hand side of Eq. (6) as

$$G \equiv 2\pi \sum_{i=1}^{m} \int_{x_{i-1}}^{x_i} n_i(x)g(x)x \, dx, \tag{9}$$

which can be interpreted as the labor demand function for the city as a whole. Then it follows from Eqs. (4) and (5) that given x_i $(i = 1, \ldots, m)$,

$$\partial G/\partial w < 0; \tag{10}$$

i.e., the total demand for labor decreases as the wage rate rises, since r_i and w are inversely related from Eq. (4) and the labor–land ratio n_i declines everywhere as w increases and r_i decreases. Thus, by applying the implicit function theorem to Eq. (6), we may express w as

$$w = w(x_1, \ldots, x_m; N) \tag{11}$$

with

$$w_i \equiv \partial w/\partial x_i = b \, \partial G/\partial x_i = 2\pi b g(x_i)x_i[n_i(x_i) - n_{i+1}(x_i)]$$
$$(i = 1, \ldots, m-1), \tag{12}$$

$$w_m \equiv \partial w/\partial x_m = b \, \partial G/\partial x_m = 2\pi b g(x_m)x_m n_m(x_m) > 0, \tag{13}$$

$$w_N \equiv \partial w/\partial N = -b < 0, \tag{14}$$

where $b = -1/(\partial G/\partial w) > 0$ from Eq. (10). In view of Eqs. (2) and (11), industry i's bid rent at the boundary x_i can be written as

$$r_i(x_i) = r_i[x_i, w(x_1, \ldots, x_m; N)] \qquad (i = 1, \ldots, m),$$

and industry $(i + 1)$'s bid rent at x_i as

$$r_{i+1}(x_i) = r_{i+1}[x_i, w(x_1, \ldots, x_m; N)] \qquad (i = 1, \ldots, m-1).$$

Now we can rewrite the dynamic adjustment process (8) as

$$\dot{x}_i = f_i\{r_i[x_i, w(x_1, \ldots, x_m; N)] - r_{i+1}[x_i, w(x_1, \ldots, x_m; N)]\}$$
$$(i = 1, \ldots, m-1), \tag{15}$$

$$\dot{x}_m = f_m\{r_m[x_m, w(x_1, \ldots, x_m; N)] - r_0\}.$$

In a small neighborhood of equilibrium, system (15) may be approximated by the linear differential equation system

$$\dot{z} = FAz, \tag{16}$$

where F is a diagonal matrix whose ith diagonal element is $f_i'(0)$ $(i = 1, \ldots, m)$, z is a column vector whose ith element is $x_i - x_i^*$ $(1 = i, \ldots, m)$ and A is an

$m \times m$ matrix

$$A \equiv \mathrm{diag}[R_i] + [T_i w_j], \tag{17}$$

which consists of a diagonal matrix $\mathrm{diag}[R_i]$ with

$$R_i \equiv r_{ix}(x_i) - r_{i+1x}(x_i) = -\left[\frac{q_i'(x_i)}{h_i(x_i)} - \frac{q_{i+1}'(x_i)}{h_{i+1}(x_i)}\right]$$
$$(i = 1,\ldots,m-1), \tag{18}$$

$$R_m \equiv r_{mx}(x_m) = -q_m'(x_m)/h_m(x_m) < 0, \tag{19}$$

and a matrix $[T_i w_j]$ with

$$T_i w_j \equiv [r_{iw}(x_i) - r_{i+1w}(x_i)]w_j = -[n_i(x_i) - n_{i+1}(x_i)]w_j$$
$$(i = 1,\ldots,m-1; \quad j = 1,\ldots,m), \tag{20}$$

$$T_m w_j \equiv r_{mw}(x_m)w_j = -n_m(x_m)w_j \qquad (j = 1,\ldots,m). \tag{21}$$

Here, all the variables in Eqs. (16)–(21) are evaluated at the equilibrium.
We now prove the following theorem.

Theorem 1.1   *In system* (15), *the equilibrium is locally stable for any positive speeds of adjustment* $f_i'(0) > 0$ $(i = 1,\ldots,m)$, *if the following condition is satisfied:*

$$q_i'(x_i^*)/h_i(x_i^*) > q_{i+1}'(x_i^*)/h_{i+1}(x_i^*) \qquad (i = 1,\ldots,m-1). \tag{22}$$

Proof   According to Arrow and McManus (1958), the linear system (16) is stable for any choice of positive diagonal matrix F, if there exists a positive diagonal matrix D such that DA is symmetric and negative definite, thus obviously quasi-negative definite. Let $D \equiv \mathrm{diag}[D_i]$ with its ith diagonal element $D_i \equiv g(x_i)x_i > 0$ $(i = 1,\ldots,m)$. Then it follows from (17) that

$$DA = \mathrm{diag}[D_i R_i] + [D_i T_i w_j] = \mathrm{diag}[D_i R_i] - [s_i s_j], \tag{23}$$

where

$$s_i \equiv \sqrt{2\pi b g(x_i) x_i [n_i(x_i) - n_{i+1}(x_i)]} \qquad (i = 1,\ldots,m-1),$$
$$s_m \equiv \sqrt{2\pi b g(x_m) x_m n_m(x_m)},$$

in view of (12), (13), (20), and (21). Since $R_i < 0$ from (18) and (22), we have the negative definite quadratic form

$$y'DAy = \sum_{i=1}^{m} D_i R_i(y_i)^2 - \left(\sum_{i=1}^{m} s_i y_i\right)^2 < 0$$

for all nonzero vectors y. Thus, DA is negative definite.   Q.E.D.

Condition (22) states that at each boundary between industrial zones, the industry located in the inner zone should have a higher ratio of marginal transport cost to land per unit of output than the industry in the outer zone at the equilibrium. In fact, this is known as the "von Thünen condition," and is equivalent to the condition that at each boundary the bid rent curve of the inner industry be steeper than that of the outer industry, i.e.,

$$r_{ix}(x_i^*) < r_{i+ix}(x_i^*) \qquad (i = 1, \ldots, m - 1) \qquad (24)$$

in view of (18). Condition (24) is actually satisfied by the bid rent curves illustrated in Fig. 1.1. It should be noted, however, that although in the literature the von Thünen condition (22) or (24) is regarded as a condition for spatial equilibrium with industries located in concentric rings,[8] it is in fact a little stronger than necessary to sustain such an equilibrium, which only requires

$$r_{ix}(x_i^*) \leqq r_{i+1x}(x_i^*) \qquad (i = 1, \ldots, m - 1). \qquad (25)$$

Our dynamic analysis shows that stability is ensured under condition (24), but not under condition (25).

1.3   Wage Adjustment

Let us turn to an alternative adjustment process, i.e., a gradual adjustment of the wage rate in the labor market. It seems natural to suppose that the wage rate will rise *or* fall, according as the excess demand for labor is positive *or* negative.[9] During the course of wage adjustment, the full employment condition (6) is not satisfied, whereas all the other variables r_i, n_i, and x_i are assumed to be fully adjusted so as to satisfy Eqs. (1), (5), and (7) instantaneously.

In order to express total labor demand as a function of the wage rate, we first note that, in view of Eqs. (2) and (8), x_i's are determined for each given value of w by solving the market rent condition of the form

$$r_i(x_i, w) = r_{i+1}(x_i, w) \qquad (i = i, \ldots, m - 1),$$

$$r_m(x_m, w) = r_0. \qquad (26)$$

Once x_i's are expressed as functions of w, we can obtain r_i's and n_i's as functions of w from Eqs. (2) and (5), and also the total demand for labor, defined

---

[8] See, e.g., Alonso (1964), Beckmann (1972), Mills (1972), Muth (1961), and Solow (1973).

[9] A similar adjustment process is implicitly assumed by Solow (1973).

in Eq. (9), as a function of w, $G = G(w)$. Then the dynamic process of wage adjustment can be formulated as

$$\dot{w} = f[G(w) - N] \tag{27}$$

with $f'(\;) > 0$, $f(0) = 0$.

Here we present the following useful lemma.

Lemma 1.1   *In system* (26), *we have*

$$dx_i/dw \gtreqless 0, \quad as \quad n_i(x_i) \gtreqless n_{i+1}(x_i) \quad (i = 1, \ldots, m - 1).$$
$$dx_m/dw < 0, \tag{28}$$

if the following condition is met:

$$q_i'(x_i)/h_i(x_i) > q_{i+1}'(x_i)/h_{i+1}(x_i) \quad (i = 1, \ldots, m - 1). \tag{29}$$

Proof   by differentiating (26) with respect to w, we find

$$[r_{ix}(x_i) - r_{i+1x}(x_i)](dx_i/dw) = -[r_{iw}(x_i) - r_{i+1w}(x_i)]$$
$$(i = 1, \ldots, m - 1), \tag{30}$$

$$r_{mx}(x_m)(dx_m/dw) = -r_{mw}(x_m).$$

Thus, (28) follows from (30) together with (3), (4), and (29).   Q.E.D.

Now we are ready to prove the following theorem.

Theorem 1.2   *According to process* (27), *the equilibrium is locally stable for any positive speed of adjustment* $f'(0) > 0$, *if the von Thünen condition* (22) *is met.*

Proof   In a small neighborhood of equilibrium, we can approximate (27) as

$$(\dot{w} \doteq w^*) = f'(0)G'(w^*)(w - w^*). \tag{31}$$

According to (31), the equilibrium is stable, since $f'(0) > 0$ and

$$G'(w^*) = 2\pi \left\{ \sum_{i=1}^{m} \int_{x_{i-1}}^{x_i} \frac{dn_i}{dw} g(x)x \, dx \right.$$
$$\left. + \sum_{i=1}^{m-1} [n_i(x_i) - n_{i+1}(x_i)]g(x_i)x_i \frac{dx_i}{dw} + n_m(x_m)x_m \frac{dx_m}{dw} \right\} < 0, \tag{32}$$

from Lemma 1.1 and $dn_i/dw < 0$, where all the variables are evaluated at the equilibrium.   Q.E.D.

1.4   Comparative Statics

Having examined the dynamic property of the model, we are now in a position to conduct a comparative static analysis. It turns out that some comparative static results can be derived directly from our stability analysis. Specifically, we can find the effects of changes in the availability of labor and land on the equilibrium wage rate w^* and boundary positions x_i^*.

First, let us consider a small change in the total amount of labor N. Under condition (22), it follows from Eqs. (6), (9), and (32) that

$$dw^*/dN = 1/G'(w^*) < 0.$$

Since $dx_i/dN = (dx_i/dw)(dw/dN)$, Lemma 1.1 leads to the following theorem.

Theorem 1.3   *Under condition (22), we have*

$$dw^*/dN < 0, \tag{33}$$

$$dx_i^*/dN \gtreqless 0, \qquad as \quad n_i(x_i^*) \gtreqless n_{i+1}(x_i^*) \qquad (i = 1,\ldots,m-1), \tag{34}$$

$$dx_m^*/dN > 0. \tag{35}$$

The economic interpretation of this result is quite straightforward. Equation (33) states that when the total supply of labor increases in the city, the equilibrium wage rate will fall. According to (34), a relatively more labor-intensive industry will expand its boundary against a relatively less labor-intensive industry at each boundary, when the total amount of labor increases. Note that this is somewhat similar to a well-known theorem by Rybczynski (1955). Finally, Eq. (35) says that the outer boundary of the city will expand with an increase in the total amount of labor available in the city.[10]

Next we shall examine the effect of a change in the availability of land for production inside the city. For this purpose, we assume a small change in g, i.e., the fraction of land devoted to production at each distance from the city center. Introduce a shift parameter v into the function g as $g(x, v)$, and assume that

$$\partial g(x, v)/\partial v \geq 0 \qquad \text{for} \quad 0 \leq x \leq x_m,$$

---

[10] Note that in the special case considered by Alessio (1973), Beckmann (1972), and Renaud (1972), where production functions are "linearly dependent" for all industries, the labor–land ratios of two successive industries are the same at each boundary and, therefore, only the outer city boundary will move outward, whereas none of the remaining boundaries will move in either direction, as the total amount of labor increases.

and that there exist nonnegative constants $\underline{x}$ and $\bar{x}$ such that $0 \leqq \underline{x} < \bar{x} \leqq x_m$ and

$$\partial g(x, v)/\partial v > 0 \qquad \text{for} \quad \underline{x} \leqq x \leqq \bar{x}.$$

This means that somewhere in the city, a greater fraction of the land area in a ring of positive width becomes available for production, as the parameter v increases.

The full employment condition (6) can then be written as

$$G(w, v) = N \tag{36}$$

in view of (9). Under condition (22), we obtain $\partial G/\partial w < 0$ from Eq. (32). We can also find $\partial G/\partial v > 0$, since in our model g appears only in Eq. (6) and, therefore, for each given value of w an increase in v shifts the labor demand function upward without affecting x_i's. Thus, under condition (22) we see $dw^*/dv = -(\partial G/\partial v)/(\partial G/\partial w) > 0$, which together with Lemma 1.1 leads to the following theorem.

Theorem 1.4   *Under condition* (22),

$$dw^*/dv > 0, \tag{37}$$

$$dx_i^*/dv \gtreqqless 0, \qquad as \quad n_i(x_i^*) \gtreqqless n_{i+1}(x_i^*) \qquad (i = 1, \ldots, m - 1), \tag{38}$$

$$dx_m^*/dv < 0. \tag{39}$$

Note that this theorem holds regardless of where we have $\partial g/\partial v > 0$ in the city. This yields the following interesting and rather counterintuitive case. Suppose that due to a change in zoning regulations, for example, a greater fraction of land becomes available for production in zone i, where industry i happens to be relatively less labor intensive at its inner boundary as well as its outer boundary, i.e., $n_{i-1}(x_{i-1}^*) > n_i(x_{i-1}^*)$ and $n_i(x_i^*) < n_{i+1}(x_i^*)$. Then, industry i will expand, rather than contract, in both inward and outward directions, as more land becomes available to that industry within the boundaries of zone i. This is because relatively more labor intensive industries should contract and release some of their labor force to the newly available land for production in order to maintain the full employment condition.

Finally, we shall consider the effect of a change in the opportunity cost of land r_0 on the equilibrium values of w and x_i's. It is noted that the opportunity cost of land can be affected by some public policies, e.g., a change in the rate of a land tax. Setting $x_i = 0$ $(i = 1, \ldots, m)$ in system (15) and differentiating it with respect to r_0, we find the relation for a small neighborhood of equilibrium,

$$Ay = e, \tag{40}$$

where y is an $m \times 1$ vector whose ith element is dx_i^*/dr_0 $(i = 1, \ldots, m)$, e is an $m \times 1$ vector with its mth element equal to unity and all other elements zero, and A is given by Eq. (17).

Now we can prove the following theorem.

Theorem 1.5   *Under condition* (22), *we obtain*

$$dw^*/dr_0 < 0, \tag{41}$$

$$dx_i^*/dr_0 \gtreqless 0, \quad as \quad n_i(x_i^*) \gtreqless n_{i+1}(x_i^*) \quad (i = 1, \ldots, m-1), \tag{42}$$

$$dx_m^*/dr_0 < 0. \tag{43}$$

Proof   By solving (40) for y, we find

$$dx_i^*/dr_0 = |A|_i/|A| = (-1)^{m+1}\left[(-1)^{m-i-1}\left(\prod_{k=1}^{m-1} R_k\right)(R_i)^{-1}T_iw_m\right]$$

$$\times \left\{\prod_{j=1}^{m} R_j + \sum_{k=1}^{m}\left[\left(\prod_{k=1}^{m} R_k\right)(R_j)^{-1}T_jw_j\right]\right\}^{-1}$$

$$(i = 1, \ldots, m-1), \tag{44}$$

where $|A|_i$ is the determinant of the matrix A with its ith column replaced by e, and $|A|$ is the determinant of A. Since $R_i > 0$ $(i = 1, \ldots, m)$ from (18), (19), and (22), and $T_jw_j \leq 0$ $(j = 1, \ldots, m)$ from (12), (13), (20), and (21), property (42) follows from (44) in view of (12) and (20). Similarly, we obtain (43) as

$$dx_m^*/dr_0 = \left\{\prod_{j=1}^{m-1} R_j + \sum_{k=1}^{m-1}\left[\left(\prod_{j=1}^{m-1} R_j\right)(R_k)^{-1}T_kw_k\right]\right\}/|A| < 0. \tag{45}$$

In order to show (41), we derive from the proof of Lemma 1.1

$$dw/dx_i = -[r_{ix}(x_i) - r_{i+1x}(x_i)]/[r_{iw}(x_i) - r_{i+1w}(x_i)] = -R_i/T_i,$$

where the last equality comes from (18) and (20). Then,

$$dw^*/dr_0 = (dw^*/dx_i^*)(dx_i^*/dr_0),$$

which is equal to (44) with T_i in the numerator replaced by $(-R_i)$ and, therefore, is negative.   Q.E.D.

Equation (41) means that the "factor price frontier" in our model is negatively sloped, as is always the case. According to Eqs. (42) and (43), a boundary between two industries inside the city will move outward *or* inward, according as the inner industry is more *or* less labor intensive than the outer industry at the boundary, whereas the city will always decrease in size, as the opportunity cost of land rises. It may be interesting to compare this result with Theorem 1.4. We find that a *decrease* in the availability of land for production inside the city and an *increase* in the opportunity cost of land in and around the city will have exactly the same effects on the wage rate and the industrial boundaries inside the city, but will have completely the opposite effects on the outer city boundary: the former leads to an expansion of the city, whereas the latter results in a contraction.

1.5   The LeChatelier Principle

Given Theorem 1.5, particularly the last property, (43), we shall further ask how the rate of change of the outer city boundary is affected when some or all of the boundaries inside the city are not allowed to move in response to a change in the opportunity cost of land, due to frictions, time lags, or regulations in land use. Specifically, we shall evaluate dx_m^*/dr_0 by increasing the number of boundries to be held constant in ascending order, i.e., from x_1^* through x_{m-1}^*.

Recalling the LeChatelier Principle as applied to economic analysis by Samuelson (1947), one might conjecture that the greater the number of boundaries to be fixed, the smaller the rate of change of the outer city boundary. In fact, defining $(dx_m^*/dr_0)_i$ as dx_m^*/dr_0 with the first i boundaries $(x_1, \ldots, x_i)$ fixed at their initial equilibrium values $(i = 1, \ldots, m - 1)$, we can prove the following theorem.

Theorem 1.6   *Under condition* (22), *we find*

$$dx_m^*/dr_0 \leqq (dx_m^*/dr_0)_1 \leqq (dx_m^*/dr_0)_2 \leqq \cdots \leqq (dx_m^*/dr_0)_{m-1} < 0, \quad (46)$$

where ( )$_i$ *means that the first i boundaries $x_1, \ldots, x_i$ are fixed at their equilibrium values $(i = 1, \ldots, m - 1)$.*

Proof   Let A_{ii} be the matrix A with its ith row and ith column deleted, y_i the vector y with its ith element deleted, e_i the vector e with its ith element deleted, A_{iijj} the matrix A with its ith and jth rows and its ith and jth columns deleted, and so on. First, by holding x_1^* constant, we obtain $A_{11}y_1 = e_1$ and $(dx_m^*/dr_0)_1 = |A_{11mm}|/|A_{11}|$, which is shown to be negative by the same reasoning as (45). Utilizing Jacobi's identity (Samuelson, 1947, p. 370).

$$|A||A_{iijj}| = |A_{ii}||A_{jj}| - |A_{ij}||A_{ji}|, \quad (47)$$

and setting $i = 1$ and $j = m$, we find

$$dx_m^*/dr_0 - (dx_m^*/dr_0)_1 = |A_{mm}|/|A| - |A_{11mm}|/|A_{11}|$$

$$= (|A_{mm}||A_{11}| - |A||A_{11mm}|)/(|A||A_{11}|)$$

$$= |A_{1m}||A_{m1}|/(|A||A_{11}|).$$

Here, under condition (22), the denominator $|A||A_{11}|$ is obviously negative, and the numerator is nonnegative as

$$|A_{1m}||A_{m1}| = |(DA)_{1m}||(DA)_{m1}|(\prod_{i=1}^{m-1} D_i)^{-1}(\prod_{i=2}^{m} D_i)^{-1} \geqq 0,$$

where D is a positive diagonal matrix with its ith diagonal element $D_i \equiv g(x_i^*)x_i^* > 0$ $(i = 1,\ldots,m)$ and, therefore, DA is symmetric in view of (12) to (21). Thus, $dx_m^*/dr_0 \leqq (dx_m^*/dr_0)_1 < 0$. Next, apply the same argument to the system $A_{1122}y_{12} = e_{12}$, which has the same desirable properties as above, and obtain $(dx_m^*/dr_0)_1 \leqq (dx_m^*/dr_0)_2 < 0$. The theorem is proved by induction.   Q.E.D.

By applying the LeChatelier Principle, we have shown that the rate of change of the outer city boundary with respect to a change in the opportunity cost of land will be greater, as more boundaries between industries inside the city are allowed to move in response, regardless of whether those boundaries between industries move inward or outward in response to such a change.[11] An immediate policy implication of this result is that land tax policies aiming at restricting or expanding the overall city size will be most (more) effective, when there are no (fewer) land-use regulations to freeze the existing locations of industries inside the city. More generally, it can be shown that the effect of a given change in *any* parameter on the outer city boundary will be greater, as more boundaries inside the city are permitted to vary in response to such a change. This result might be regarded as an example to illustrate how important a proper coordination of land use policies will be for their effectiveness in influencing the size and configurations of a city.

REFERENCES

Alessio, F. J. (1973). A neoclassical land use model: The influence of externalities. *Swedish Journal of Economics* **75**, 414–419.
Alonso, W. (1964). *Location and Land Use*. Harvard Univ. Press, Cambridge, Massachusetts.
Arrow, K. J., and McManus, M. (1958). A note on dynamic stability. *Econometrica* **26**, 297–305.
Beckmann, M. J. (1972). Von Thünen revisited: A neoclassical land use model. *Swedish Journal of Economics* **74**, 1–7.
Diewert, W. E. (1974). "Applications of duality theory. In *Frontiers of Quantitative Economics* (M. Intriligator and D. Kendrick, eds.), Vol. II, pp. 106–171. North-Holland Publ., Amsterdam.

[11] This is an exact counterpart of Samuelson's well-known statement about the LeChatelier Principle: "A lengthening of the time period so as to permit new factors to be varied will result in greater changes in the factor whose price has changed, regardless of whether the factors permitted to vary are complementary or competitive with the one whose price has changed" (Samuelson, 1947, pp. 38–39).

Mills, E. S. (1967). An aggregative model of resource allocation in a metropolitan area. *American Economic Review* **57**, 197–210.

Mills, E. S. (1972). *Studies in the Structure of the Urban Economy*. Johns Hopkins Press, Baltimore, Maryland.

Miyao, T. (1977). Some dynamic and comparative static properties of a spatial model of production. *Review of Economic Studies* **44**, 321–327.

Miyao, T. (1980). An application of the LeChatelier principle in location theory. *Journal of Urban Economics* **7**, 168–174.

Muth, R. F. (1961). Economic change and rural–urban land conversions. *Econometrica* **29**, 1–23.

Renaud, B. M. (1972). On a neoclassical model of land use. *Swedish Journal of Economics* **74**, 400–404.

Rybczynski, T. M. (1955). Factor endowment and relative commodity prices. *Economica* **22**, 336–341.

Samuelson, P. A. (1947). *Foundations of Economic Analysis*. Harvard Univ. Press, Cambridge, Massachusetts.

Samuelson, P. A. (1960). An extension of the LeChatelier principle. *Econometrica* **28**, 368–379.

Shephard, R. (1953). *Cost and Production Functions*. Princeton Univ. Press, Princeton, New Jersey.

Shephard, R. (1970). *Theory of Cost and Production Functions*. Princeton Univ. Press, Princeton, New Jersey.

Solow, R. M. (1973). On equilibrium models of urban location. In *Essays in Modern Economics* (M. Parkin, ed.), pp. 2–16. Longman Group, London.

Varian, H. R. (1978). *Microeconomic Analysis*. Norton, New York.

CHAPTER 2

Residential Location

In this chapter we turn to the residential aspect of the urban economy.[1] We set up a residential location model with many "classes" of households locating themselves in a monocentric city. Each class consists of a given number of households which are identical in terms of utility functions, income levels, and transport (commuting) cost functions, whereas households in different classes possess different utility functions, different income levels, and/or different transport cost functions.

We find conditions for spatial equilibrium, and examine the stability property of the equilibrium by introducing a dynamic adjustment process of boundaries between residential classes. It is shown that the equilibrium is dynamically stable under the assumptions that (1) at each boundary between two classes the inner class has a higher ratio of marginal transport cost to residential land per household than the outer class, and that (2) land is a "non-Giffen" good for all households in the sense that the individual demand curve for land is always downward sloping. In the case of identical utility functions and identical transport cost functions for all classes, these two conditions are satisfied, if (1) richer classes live farther from the city center and (2) land is a normal good for all classes in the sense that the individual demand for land increases with income.

Given the stability conditions, we conduct a comparative static analysis to investigate the effect of changes in various parameters on the equilibrium positions of residential boundaries. It is proved that all the boundary posi-

---

[1] This chapter is a slightly generalized version of Miyao (1975).

tions will move inward, as the opportunity cost of land rises or as the rate of a residential land tax increases. Furthermore, all the boundaries are shown to expand with an increased income or size of the household class living closest to the city center. This means that in the case of identical utility functions and identical transport cost functions for all classes, as the income or the size of the poorest class increases, all the other classes will move farther away from the city center.

2.1 The Residential Model

Let us consider a circular city with a business district at the center, called the central business district (CBD). Since all employment is assumed to take place at the CBD, every resident must commute to the center, wherever he lives in the city. As in the previous chapter, it is assumed that roads are laid out radially and densely, and that transport cost (commuting cost) is a sole function of distance to travel.[2] Each household receives a fixed amount of income and spends it on a consumption good, residential land, and transportation. There are a finite number of household classes with different utility functions, different income levels, and/or different transport cost functions, whereas each class consists of a given number of households which are identical in terms of income, utility, and transport cost functions. In equilibrium, households are located in concentric rings, or zones, around the CBD, and each zone is occupied by a single class of households. Both zones and classes are numbered in ascending order of distance from the CBD so that zone i corresponds to class i.

First, focus on a household in class i, which is located at a distance x from the CBD. The household maximizes its utility U_i, which depends on d_i, the amount of a consumption good, and h_i, the amount of residential land (space),

$$U_i(x) = U_i[d_i(x), h_i(x)],$$

subject to its budget constraint

$$d_i(x) + r(x)h_i(x) = w_i - q_i(x),$$

where the price of the consumption good, being given exogenously, is

---

[2] Here we take account of "out-of-pocket" cost only and not of congestion cost, which will be dealt with in Part 4. The basic assumptions made in this chapter are the same as those of Solow (1973). Also see Alonso (1964), Beckmann (1969, 1973), Hartwick *et al.*, (1976), Montesano (1972), Muth (1969), and Wheaton (1974, 1976).

normalized as unity, $r(x)$ is the rental price of land at x, w_i is class i's income per household, and $q_i(x)$ is class i's transport (commuting) cost per household, which is assumed to be a strictly increasing function of distance x. Then, we can express the maximized level of utility in terms of the indirect utility function[3]

$$V_i(x) = V_i[r(x), w_i - q_i(x)].$$

In equilibrium, all households in class i must achieve the same utility level, say u_i, which is independent of x, because, otherwise, some household could improve its welfare by moving to a better location in the city. Thus, we have

$$V_i[r_i(x), w_i - q_i(x)] = u_i \qquad (i = 1, \ldots, m), \tag{1}$$

which defines class i's bid rent $r_i(x)$, i.e., the maximum rent that households in class i are willing to pay for each location x, while maintaining the utility level u_i. Also, for zone i to be occupied by class i, it is necessary that class i's bid rent be higher than or equal to that of any other class everywhere in zone i:

$$r_i(x) \geqq r_j(x) \qquad \text{for all} \quad j = 1, \ldots, m, \quad \text{for} \quad x_{i-1} \leqq x \leqq x_i,$$

where x_{i-1} and x_i are the inner and outer boundaries (radii) of zone i, respectively.[4] This corresponds to the fact that class i's bid rents become the prevailing market rents inside zone i.

Defining $V_{ir} \equiv \partial V_i/\partial r_i$ and $V_{iw} \equiv \partial V_i/\partial w_i$, we recall the well-known properties of the indirect utility function to note

$$V_{ir} = -h_i V_{iw} \qquad (i = 1, \ldots, m), \tag{2}$$

which is negative, as we naturally assume that the marginal utility of income, V_{iw}, is positive, and the individual demand for land, h_i, is also positive. Then, we can solve Eq. (1) for r_i as

$$r_i(x) = r_i[w_i - q_i(x), u_i] \qquad \text{for} \quad x_{i-1} \leqq x \leqq x_i \quad (i = 1, \ldots, m). \tag{3}$$

with

$$r_{iw}(x) \equiv \partial r_i/\partial w_i = -V_{iw}/V_{ir} > 0, \tag{4}$$

$$r_{iu}(x) \equiv \partial r_i/\partial u_i = 1/V_{ir} < 0. \tag{5}$$

---

[3] See, e.g., Samuelson (1947) and Varian (1978) on various properties of the indirect utility function.

[4] It remains to be investigated what assumptions on incomes and preferences will actually lead to such a segregated residential pattern. For further discussions on this question, see Mills (1972, p. 71) and Muth (1969, p. 29).

It also follows from (2) that class i's demand for land per household is a function of r_i and $w_i - q_i$ as

$$h_i(x) = -V_{ir}/V_{iw} \equiv h_i[r_i(x), w_i - q_i(x)] \qquad \text{for} \quad x_{i-1} \leqq x \leqq x_i$$

$$(i = 1, \ldots, m) \qquad (6)$$

At each distance x from the CBD, a fraction g of the total area of a thin ring with inner radius x and outer radius $x + dx$ is assumed to be available for residential use, where $g(x) > 0$ for all x. Noting that $1/h_i(x)$ is the density of households in class i at x, the number of households in class i located in the thin ring at x is equal to $[1/h_i(x)]g(x)2\pi x \, dx$. Then the "full accommodation condition" stating that all households in class i should be housed in zone i may be written as

$$2\pi \int_{x_{i-1}}^{x_i} \frac{g(x)x}{h_i[r_i(x), w_i - q_i(x)]} \, dx = N_i \qquad (i = 1, \ldots, m), \qquad (7)$$

where the total number of households in class i, N_i, is given as a positive constant.

Since class i's bid rent function in zone i represents the segment of the market rent function between x_{i-1} and x_i, the "market rent condition" that the market rent function should be continuous at every boundary is expressed as

$$r_i(x_i) = r_{i+1}(x_i) \qquad (i = 1, \ldots, m - 1),$$

$$r_m(x_m) = r_0, \qquad (8)$$

where r_0 is the opportunity cost of land, e.g., agricultural land rent, which is given exogenously.

In our model, conditions (1) and (6)–(8) can determine the equilibrium functional forms of r_i and $h_i(i = 1, \ldots, m)$ and the equilibrium values of u_i and x_i $(i = 1, \ldots, m)$ simultaneously. Here, the existence of such an equilibrium is simply assumed in order to focus on dynamics and comparative statics.

2.2 Dynamic Stability

Suppose there is a small displacement of the boundaries from their equilibrium positions and the market rent function is no longer continuous at arbitrarily given boundary positions. In other words, the market rent condition (8) is not necessarily met for arbitrarily given values of x_i's, provided that all the other conditions (1), (6), and (7) are satisfied. We now introduce a dynamic adjustment process of boundary positions, as in the previous chapter; i.e., each boundary will be adjusted gradually through time in such a way as to move outward *or* inward, according as the bid rent of the inner class is higher *or* lower than that of the outer class at each boundary; and the rate of change of a boundary with respect to time will diminish, as the rent difference narrows at the boundary in question. This adjustment process implies that it takes time for a class of households willing to pay higher rents than another class of households actually living at a boundary to outbid them and move across the existing boundary, partly because such factors as moving cost, uncertainty, lack of information, and interdependency among households make household movement rather sluggish, and partly because the movement of a residential boundary involves the removal and construction of durable objects, e.g., houses, roads, and other public facilities, which can be changed only gradually over time. The adjustment process may then be formally written

$$\dot{x}_i = f_i[r_i(x_i) - r_{i+1}(x_i)] \qquad (i = 1, \ldots, m-1),$$
$$\dot{x}_m = f_m[r_m(x_m) - r_0], \tag{9}$$

where the dot indicates differentiation with respect to time, and

$$f_i'(\ \ ) > 0, \qquad f_i(0) = 0 \qquad (i - 1, \ldots, m).$$

Let us define the left-hand side of Eq. (7) by G_i ($i = 1, \ldots, m$). In view of Eq. (3), G_i is a function of u_i, x_{i-1}, x_i and w_i,

$$G_i(u_i, x_{i-1}, x_i, w_i) = 2\pi \int_{x_{i-1}}^{x_i} \frac{g(x)x}{h_i\{r_i[w_i - q_i(x), u_i], w_i - q_i(x)\}} \, dx$$
$$(i = 1, \ldots, m). \tag{10}$$

By solving the full accommodation condition (7) or

$$G_i(u_i, x_{i-1}, x_i, w_i) = N_i \qquad (i = 1, \ldots, m) \tag{11}$$

for u_i, we can determine u_i as a function of x_{i-1}, x_i, w_i, and N_i,

$$u_i = u_i(x_{i-1}, x_i, w_i, N_i) \qquad (i = 1, \ldots, m). \tag{12}$$

Then, we prove the following lemma.

Lemma 2.1   *From Eq. (12), we find*

$$u_{ii-1} \equiv \frac{\partial u_i}{\partial x_{i-1}} < 0, \qquad u_{ii} \equiv \frac{\partial u_i}{\partial x_i} > 0, \qquad u_{iw} \equiv \frac{\partial u_i}{\partial w_i} > 0, \qquad u_{iN} \equiv \frac{\partial u_i}{\partial N_i} < 0,$$
$$(i = 1, \ldots, m), \tag{13}$$

if residential land is a "non-Giffen" good for all households, i.e.,

$$\partial h_i[r_i(x), w_i - q_i(x)]/\partial r_i < 0 \qquad \text{for} \quad x_{i-1} \leqq x \leqq x_i$$

$$(i = 1, \ldots, m). \qquad (14)$$

Proof   First, note from (10) that

$$G_{iu} \equiv \frac{\partial G_i}{\partial u_i} = -2\pi \int_{x_{i-1}}^{x_i} \frac{g(x)x}{(h_i)^2} \frac{\partial h_i}{\partial r_i} r_{iu} \, dx < 0, \qquad (15)$$

in view of (5) and (14). Thus, from (10) and (11)

$$u_{ii-1} = 2\pi g(x_{i-1})x_{i-1}/G_{iu}h_i(x_{i-1}) < 0, \qquad u_{ii} = -2\pi g(x_i)x_i/G_{iu}h_i(x_i) > 0,$$

$$u_{iw} = \frac{2\pi}{G_{iu}} \int_{x_{i-1}}^{x_i} \frac{g(x)x}{(h_i)^2} \left[\frac{\partial h_i}{\partial r_i} r_{iw} + \frac{\partial h_i}{\partial w_i} \right] dx > 0,$$

where the expression within the brackets represents the effect of an increase in w_i on h_i with u_i unchanged, i.e., the substitution effect which is always negative; and

$$u_{iN} = 1/G_{iu} < 0. \quad \text{Q.E.D.}$$

In Lemma 2.1, we assume residential land to be a non-Giffen good in the sense that the individual (Marshallian) demand curve is always downward sloping. This is the case if land is a normal good, i.e., $\partial h_i/\partial w_i > 0$, because in this case the substitution and income effects are reinforcing and together leading to $\partial h_i/\partial r_i < 0$. However, land can be a non-Giffen good even when it is an inferior good, in which case the substitution and income effects are offsetting, and land *is* non-Giffen good if the substitution effect dominates the income effect.

Taking Eqs. (3) and (12) into account, we rewrite the dynamic process (9) as

$$\dot{x}_i = f_i\{r_i[w_i - q_i(x_i), u_i(x_{i-1}, x_i, w_i, N_i)]$$

$$-r_{i+1}[w_{i+1} - q_{i+1}(x_i), u_{i+1}(x_i, x_{i+1}, w_{i+1}, N_{i+1})]\}$$

$$(i = 1, \ldots, m - 1), \qquad (16)$$

$$\dot{x}_m = f_m\{r_m[w_m - q_m(x_m), u_m(x_{m-1}, x_m, w_m, N_m)] - r_0\}.$$

Then it is clear that the stationary equilibrium condition $\dot{x}_i = 0, (i = 1, \ldots, m)$ will give the equilibrium positions of the boundaries x_i^* $(i = 1, \ldots, m)$. For a small displacement from equilibrium, the dynamic system (16) can be approximated by the linear differential equation system

$$\dot{z} = FAz, \qquad (17)$$

where F is a diagonal matrix whose ith diagonal element is $f_i'(0)$ $(i = 1, \ldots, m)$, and z is a column vector whose ith element is $x_i - x_i^*$ $(i = 1, \ldots, m)$; and

$$
A = \begin{bmatrix}
a_{11} & a_{12} & 0 & 0 & \cdots & & 0 \\
a_{21} & a_{22} & a_{23} & 0 & & & \cdot \\
0 & a_{32} & a_{33} & \cdot & & & 0 \\
& & & \cdot & \cdot & & \\
0 & 0 & \cdot & & \cdot & & \cdot \\
& & & & \cdot & a_{m-1m} & \\
\cdot & & & & \cdot & & \\
0 & \cdot & \cdots & 0 & a_{mm-1} & a_{mm}
\end{bmatrix}, \tag{18}
$$

where

$$
\begin{aligned}
a_{ii} &= -\left[r_{iw}(x_i^*) q_i'(x_i^*) - r_{i+1w}(x_i^*) q_{i+1}'(x_i^*) \right] + r_{iu}(x_i^*) u_{ii} \\
& \quad - r_{i+1u}(x_i^*) u_{i \mid 1i} \qquad (i = 1, \ldots, m - 1), \\
a_{mm} &= -r_{mw}(x_m^*) q_m'(x_m^*) + r_{mu}(x_m^*) u_{mm}, \\
a_{ii-1} &= r_{iu}(x_i^*) u_{ii-1} \qquad (i - 2, \ldots, m), \\
a_{ii+1} &= -r_{i+1u}(x_i^*) u_{i+1i+1} \qquad (i = 1, \ldots, m - 1),
\end{aligned} \tag{19}
$$

with all the variables being evaluated at the equilibrium.

Now we prove the following theorem.

Theorem 2.1 *In system (16), the equilibrium is locally stable for any position speeds of adjustment* $f_i'(0) > 0$ $(i = 1, \ldots, m)$ *if at each boundary between two classes of households the inner class has a higher ratio of marginal transport cost to land per household than the outer class, i.e.,*[5]

$$
q_i'(x_i^*)/h_i(x_i^*) > q_{i+1}'(x_i^*)/h_{i+1}(x_i^*) \qquad (i = 1, \ldots, m - 1), \tag{20}
$$

and land is a non-Giffen good for all classes, i.e.,

$$
\partial h_i [r_i(x), w_i - q_i(x)]/\partial r_i < 0 \qquad \text{for} \quad x_{i-1}^* \leqq x \leqq x_i^* \qquad (i = 1, \ldots, m). \tag{21}
$$

Proof First, from Eqs. (4) and (6) together with (20), we find

$$
r_{iw}(x_i) q_i'(x_i) - r_{i+1w}(x_i) q_{i+1}'(x_i) = \frac{q_i'(x_i)}{h_i(x_i)} - \frac{q_{i+1}'(x_i)}{h_{i+1}(x_i)} > 0 \tag{22}
$$

---

[5] This might be called the "von Thünen condition," as mentioned in Chapter 1.

at the equilibrium. Thus, it follows from (4), (5), (13), and (20)–(22) that

$$a_{ii} < 0 \qquad (i = 1, \ldots, m),$$

$$a_{ii-1} > 0 \qquad (i = 2, \ldots, m), \qquad (23)$$

$$a_{ii+1} > 0 \qquad (i = 1, \ldots, m-1).$$

According to McKenzie (1960), the equilibrium is stable for any positive diagonal matrix F in the linear system (17), if A has negative diagonal elements and is "quasi-dominant diagonal," i.e., there exist positive constants c_i $(i = 1, \ldots, m)$ such that

$$c_i|a_{ii}| > \sum_{k \neq i} c_k|a_{ik}| \qquad (i = 1, \ldots, m).$$

Under the present conditions, we can actually find $c_i > 0$ $(i = 1, \ldots, m)$ such that

$$-c_1 a_{11} > c_2 a_{12}, \qquad (24)$$

$$-c_i a_{ii} > c_{i-1} a_{ii-1} + c_{i+1} a_{ii+1} \qquad (i = 2, \ldots, m-1), \qquad (25)$$

$$-c_m a_{mm} > c_{m-1} a_{mm-1}. \qquad (26)$$

Let $c_1 \equiv 1$ and

$$c_i \equiv \prod_{k=2}^{i} \left(-\frac{u_{kk-1}}{u_{kk}} \right) \qquad (i = 2, \ldots, m),$$

where it is obvious that $c_i > 0$ $(i = 1, \ldots, m)$ from Lemma 2.1. Now, (24) follows from

$$r_{1w}(x_1)q_1'(x_1) - r_{2w}(x_1)q_2'(x_1) - r_{1u}(x_1)u_{11} + r_{2u}(x_1)u_{21} > (u_{21}/u_{22})r_{2u}(x_1)u_{22}$$

in view of (22). We also find (25) as

$$\prod_{k=2}^{i} \left(-\frac{u_{kk-1}}{u_{kk}} \right) \left[r_{iw}(x_i)q_i'(x_i) - r_{i+1w}(x_i)q_{i+1}'(x_i) - r_{iu}(x_i)u_{ii} \right.$$

$$+ r_{i+1u}(x_i)u_{i+1i} \Big] > \prod_{k=2}^{i-1} \left(-\frac{u_{kk-1}}{u_{kk}} \right) r_{iu}(x_i)u_{ii-1}$$

$$+ \prod_{k=2}^{i+1} \left(-\frac{u_{kk-1}}{u_{kk}} \right) \left[-r_{i+1u}(x_i)u_{i+1i+1} \right] \qquad (i = 2, \ldots, m-1),$$

where the preceding inequality comes from the equivalent expression

$$(-u_{ii-1}/u_{ii})\left[r_{iw}(x_i)q_i'(x_i) - r_{i+1w}(x_i)q_{i+1}'(x_i) - r_{iu}(x_i)u_{ii} \right.$$

$$+ r_{i+1u}(x_i)u_{i+1i} \Big] > r_{iu}(x_i)u_{ii-1} + (u_{ii-1}/u_{ii})(u_{i+1i}/u_{i+1i+1})$$

$$\times \left[-r_{i+1u}(x_i)u_{i+1i+1} \right] \qquad (i = 2, \ldots, m-1).$$

Finally, (26) is obtained from

$$(-u_{mm-1}/u_{mm})[r_{mw}(x_m)q'_m(x_m) - r_{mu}(x_m)u_{mm}] > r_{mu}(x_m)u_{mm-1}. \quad Q.E.D.$$

Having proved stability under general assumptions, we may consider a special case in which both utility functions and transport cost functions are identical for all classes of households. In this case, the "von Thünen condition" (20) reduces to

$$h[r_i(x_i^*), w_i - q(x_i^*)] < h[r_{i+1}(x_i^*), w_{i+1} - q(x_i^*)]$$
$$(i = 1, \ldots, m-1), \qquad (27)$$

which, in view of the fact that $r_i(x_i^*) = r_{i+1}(x_i^*)$, is implied by $w_i < w_{i+1}$ $(i = 1, \ldots, m-1)$, provided that land is a (strongly) normal good for all classes,

$$\partial h[r_i(x), w_i - q(x)]/\partial w_i > 0 \quad \text{for} \quad x_{i-1}^* \leqq x \leqq x_i^* \quad (i = 1, \ldots, m).$$

Incidentally, this normality assumption also ensures condition (21). Thus, we have found the following proposition.

Theorem 2.2 *In the case of identical utility functions and identical transport cost functions for all classes of households, the equilibrium is locally stable for any positive speeds of adjustment, if higher income classes live farther from the CBD, i.e.,*

$$w_i < w_{i+1} \quad (i = 1, \ldots, m), \qquad (28)$$

and land is a normal good for all households, i.e.,

$$\partial h[r_i(x), w_i - q(x)]/\partial w_i > 0 \quad \text{for} \quad x_{i-1}^* \leqq x \leqq x_i^* \quad (i = 1, \ldots, m).$$
$$(29)$$

It should be noted that in the literature (e.g., Beckmann, 1969, 1973; Mills, 1972; Montesano, 1972; Muth, 1969; Solow, 1973) condition (28) has been derived, not as a stability condition, but as an equilibrium condition. As pointed out in Chapter 1, however, condition (20) or (28) is a little stronger than necessary to sustain an equilibrium with different classes of households located in concentric rings.

In order to gain more insight, we assume utility functions to be Cobb–Douglas,

$$U_i(d_i, h_i) = (d_i)^{a_i}(h_i)^{b_i} \quad (i = 1, \ldots, m), \qquad (30)$$

and transport cost functions to be linear,

$$q_i(x) = q_i x \quad (i = 1, \ldots, m). \qquad (31)$$

It is a well-known property of the Cobb–Douglas utility function that

$$r_i(x)h_i(x) = \beta_i(w_i - q_i x),$$

where $\beta_i \equiv b_i/(a_i + b_i)$, i.e., the proportion of net income spent on land. Then, the von Thünen condition (20) reduces to

$$\beta_i[(w_i/q_i) - x_i^*] < \beta_{i+1}[(w_{i+1}/q_{i+1}) - x_i^*] \qquad (i = 1,\ldots,m-1), \quad (32)$$

which is implied by the following two conditions together,

$$\beta_i \leq \beta_{i+1}, \qquad w_i/q_i \leq w_{i+1}/q_{i+1} \qquad (i = 1,\ldots,m-1),$$

with strict inequality holding for either or both of these conditions. Condition (21) is also ensured in this case, because Cobb–Douglas means that every good is normal. Thus, we have proved the following theorem.

Theorem 2.3 *In the case of Cobb–Douglas utility functions* (30) *and linear transport cost functions* (31), *the equilibrium is locally stable for any positive speeds of adjustment, if either*

$$b_i/(a_i + b_i) < b_{i+1}/(a_{i+1} + b_{i+1}), \qquad w_i/q_i \leq w_{i+1}/q_{i+1}, \quad (33)$$

or

$$b_i/(a_i + b_i) \leq b_{i+1}/(a_{i+1} + b_{i+1}), \qquad w_i/q_i < w_{i+1}/q_{i+1}, \quad (34)$$

for each i *(*$i = 1,\ldots,m-1$*), where* $b_i/(a_i + b_i)$ *is the proportion of net income spent on land and* w_i/q_i *is the ratio of income to unit transport cost per unit distance per household in class* i *(*$i = 1,\ldots,m$*).*[6]

2.3 Comparative Statics

In equilibrium with $\dot{x}_i = 0$ ($i = 1,\ldots,m$), system (16) yields

$$r_i[w_i - q_i(x_i), u_i(x_{i-1}, x_i, w_i, N_i)] - r_{i+1}[w_{i+1} - q_{i+1}(x_i),$$

$$u_{i+1}(x_i, x_{i+1}, w_{i+1}, N_{i+1})] = 0 \qquad (i = 1,\ldots,m-1), \quad (35)$$

$$r_m[w_m - q_m(x_m), u_m(x_{m-1}, x_m, w_m, N_m)] - r_0 = 0,$$

which can be solved for x's as functions of the exogenous parameters r_0, w_i, and N_i ($i = 1,\ldots,m$). Now we examine how the equilibrium boundary positions will move with small changes in those parameters.

---

[6] It follows that the richer class can live closer to the CBD, i.e., $w_i > w_{i+1}$, if they have disproportionately high unit transport costs so that $w_i/q_i < w_{i+1}/q_{i+1}$.

First, we consider a small change in the opportunity cost of land r_0. From system (35) we find

$$Ax_r = e_m, \tag{36}$$

where A is defined as in Eq. (18), x_r is a column vector whose ith element is dx_i^*/dr_0 ($i = 1, \ldots, m$), and e_m is a unit column vector with its mth element equal to unity and all other elements zero. Then we can show the following property.

Theorem 2.4 *If the stability conditions* (20) *and* (21) *are satisfied, we have*

$$dx_i^*/dr_0 < 0 \qquad (i = 1, \ldots, m), \tag{37}$$

i.e., every boundary position x_i^* ($i = 1, \ldots, m$) will move inward, as the opportunity cost of land r_0 increases.

Proof As can be seen in the proof of Theorem 2.1, A is an indecomposable and quasi-dominant diagonal matrix with negative diagonal and nonnegative off-diagonal elements. Then, it follows that A^{-1} exists and $A^{-1} < 0$, i.e., all the elements of A^{-1} are strictly negative (see Quirk and Saposnik, 1968, p. 212). Thus, from (36) we find $x_r = A^{-1}e_m < 0$. Q.E.D.

Consider a proportional tax levied on residential land in the city. It can be shown that the effect of an increase in such a tax is just the same as that of an increase in the opportunity cost of land, provided that the initial level of the opportunity cost is positive. To prove this, define $\rho_i \equiv (1 + \tau)r_i$ with tax rate τ, and obtain $V_i[\rho_i(x), w_i - q_i(x)] = u_i$ in place of Eq. (1), and

$$\rho_i(x_i) = \rho_{i+1}(x_i) \qquad (i = 1, \ldots, m - 1),$$
$$\rho_m(x_m) = (1 + \tau)r_0, \tag{38}$$

where $\rho_i(\ )$ has exactly the same properties as $r_i(\ )$. Note that the "effective" cost of land for residential use at the outer city boundary is now $(1 + \tau)r_0$, because land will be taxed if it is converted into residential land. From this it follows that an increase in τ will have the same effects on the boundary positions as an increase in r_0, if $r_0 > 0$. If $r_0 = 0$, however, this kind of tax is neutral in the sense that it has no effect on the equilibrium boundary positions, as has been pointed out in the literature (see Alonso, 1964, p. 116). In view of Theorem 2.4, we have proved the following.

Theorem 2.5 *If conditions* (20) *and* (21) *are met, then*

$$dx_i^*/d\tau < 0 \qquad when \quad r_0 > 0 \qquad (i = 1, \ldots, m), \tag{39}$$

and

$$dx_i^*/d\tau = 0 \qquad when \quad r_0 = 0 \qquad (i = 1, \ldots, m). \tag{40}$$

Next, we can easily determine the effect of a small change in the number of households in the class located closest to the CBD, i.e., a small change in N_1. Differentiation of system (35) with respect to N_1 gives $Ax_N = s$, where x_N is a column vector with its ith element dx_i^*/dN_1 $(i = 1,\ldots,m)$ and s is a column vector with its first element $s_1 \equiv -r_{1u}(x_i^*)u_{1N}$, and all other elements zero, $s_i \equiv 0$ $(i = 2,\ldots,m)$. Since $r_{1u} < 0$ and $u_{1N} < 0$ from Eqs. (5) and (13) with condition (14) or (21), we find $x_N = A^{-1}s > 0$, given conditions (20) and (21). Thus, we have proved the following theorem.

Theorem 2.6 *If conditions* (20) *and* (21) *are satisfied,*

$$dx_i^*/dN_1 > 0 \qquad (i = 1,\ldots,m), \tag{41}$$

i.e., every boundary position will move outward with an increase in the number of households in the class located closest to the CBD.

Finally, we shall investigate the effect of a small change in income per household in the class living closest to the CBD, i.e., a change in w_1. It turns out that in this case we need to strengthen condition (21) slightly by assuming that land is a normal good for all households in class 1,

$$\partial h_1[r_1(x), w_1 - q_1(x)]/\partial w_1 > 0 \qquad \text{for} \quad x_0 \leqq x \leqq x_1^*, \tag{42}$$

in addition to the assumption that land is a non-Giffen good for all classes. By differentiating system (35) with respect to w_1, we find $Ax_w = k$, where x_w is a column vector with its ith element dx_i^*/dw_1 $(i = 1,\ldots,m)$ and k is a column vector with its first element $k_1 \equiv -[r_{1w}(x_1^*) + r_{1u}(x_1^*)u_{1w}]$, and all other elements zero, $k_i \equiv 0$ $(i = 2,\ldots,m)$. Because of the definition of u_{1w} given in the proof of Lemma 2.1, we obtain

$$k_1 = -\left[r_{1w}(x_1^*) - r_{1u}(x_1^*) \frac{\displaystyle\int_{x_0}^{x_1^*} \frac{g(x)x}{(h_1)^2}\left[\frac{\partial h_1}{\partial r_1}r_{1w}(x) + \frac{\partial h_1}{\partial w_1}\right]dx}{\displaystyle\int_{x_0}^{x_1^*} \frac{g(x)x}{(h_1)^2}\frac{\partial h_1}{\partial r_1}r_{1u}(x)\,dx} \right]$$

$$< -r_{1u}(x_1^*)\left[\frac{r_{1w}(x_1^*)}{r_{1u}(x_1^*)} - \frac{\displaystyle\int_{x_0}^{x_1^*} \frac{g(x)x}{(h_1)^2}\frac{\partial h_1}{\partial r_1}r_{1w}(x)\,dx}{\displaystyle\int_{x_0}^{x_1^*} \frac{g(x)x}{(h_1)^2}\frac{\partial h_1}{\partial r_1}r_{1u}(x)\,dx} \right], \tag{43}$$

provided that conditions (21) and (42) are met. The latter expression in Eq. (43) is negative because

$$-r_{1w}(x_1^*)/r_{1u}(x_1^*) > -r_{1w}(x)/r_{1u}(x) \qquad \text{for} \quad x_0 \leqq x < x_1^*,$$

which follows from

$$\frac{\partial[-r_{1w}(x)/r_{1u}(x)]}{\partial x} = \frac{\partial V_{1w}(x)}{\partial x} = \left[\frac{\partial V_{1w}}{\partial w_1} + \frac{\partial V_{1w}}{\partial r_1} r_{1w}(x)\right][-q_1'(x)]$$

$$= \left(\frac{\partial V_{1w}}{\partial w_1} - \frac{\partial V_{1r}}{\partial w_1}\frac{V_{1w}}{V_{1r}}\right)[-q_1'(x)]$$

$$= \frac{\partial(-V_{1w}/V_{1r})}{\partial w_1} V_{1r}q_1'(x) = \frac{\partial(1/h_1)}{\partial w_1} V_{1r}q_1'(x) > 0,$$

in view of Eqs. (1)–(6) as well as condition (42). Then, we have $x_w = A^{-1}k > 0$, given conditions (20), (21), and (42). Thus, the following theorem has been proved.

Theorem 2.7 *If the stability conditions (20) and (21), along with the normality condition (42), are satisfied, then*

$$dx_i^*/dw_1 > 0 \qquad (i = 1,\ldots,m), \tag{44}$$

i.e., every boundary position will move out with an increase in income per household in the class living closest to the CBD.

Finally, we should note that in the special case of identical utility functions and identical transport cost functions for all classes, the stability conditions (28) and (29) together imply conditions (20), (21), and (42). From Theorems 2.6 and 2.7, therefore, we can derive the following corollary.

Theorem 2.8 *If, in the case of identical utility functions and identical transport cost functions for all classes, the stability conditions (28) and (29) are satisfied, then we have*

$$dx_i^*/dN_1 > 0 \qquad dx_i^*/dw_1 > 0 \qquad (i = 1,\ldots,m), \tag{45}$$

i.e., every boundary position will move out with an increase in the number of households or in income per household in the poorest class.[7]

REFERENCES

Alonso, W. (1964). *Location and Land Use*. Harvard Univ. Press, Cambridge, Massachusetts.
Beckmann, M. J. (1969). On the distribution of urban rent and residential density. *Journal of Economic Theory* **1**, 60–67.

[7] For related and more general comparative static results, see Hartwick *et al.* (1976) and Wheaton (1976). Most of their results, however, are based on the assumption that land is a normal good and not just a non-Giffen good.

Beckmann, M. J. (1973). Equilibrium models of residential land use. *Regional and Urban Economics* **3**, 361–368.

Hartwick, J., Schweizer, U., and Varaiya, P. (1976). Comparative statics of a residential economy with several classes. *Journal of Economic Theory* **13**, 396–413.

McKenzie, L. W. (1960). Matrices with dominant diagonals and economic theory. In *Proceedings of a Symposium on Mathematical Methods in the Social Science* pp. 277–292. Stanford Univ. Press, Stanford, California.

Mills, E. S. (1972). *Urban Economics*, 1st ed. Scott, Foresman, Glenview, Illinois.

Miyao, T. (1975). Dynamics and comparative statics in the theory of residential location. *Journal of Economic Theory* **11**, 133–146.

Montesano, A. (1972). A restatement of Beckmann's model on the distribution of urban rent and residential density. *Journal of Economic Theory* **4**, 329–354.

Muth, R. F. (1969). *Cities and Housing*. Univ. of Chicago Press, Chicago.

Quirk, J., and Saposnik, R. (1968). *Introduction to General Equilibrium Theory and Welfare Economics*. McGraw-Hill, New York.

Samuelson, P. A. (1947). *Foundations of Economic Analysis*. Harvard Univ. Press, Cambridge, Massachusetts.

Solow, R. M. (1973). On equilibrium models of urban location. In *Essays in Modern Economics* (M. Parkin, ed.), pp. 2–16. Longman Group, London.

Varian, H. R. (1978). *Microeconomic Analysis*. Norton, New York.

Wheaton, W. C. (1974). A comparative static analysis of urban spatial structure. *Journal of Economic Theory* **9**, 223–237.

Wheaton, W. C. (1976). On the optimal distribution of income among cities. *Journal of Urban Economics* **3**, 31–44.

CHAPTER **3**

Interaction of Industrial
and Residential Locations

In the previous chapters we have dealt with industrial location and residential location separately, while paying little attention to the interaction of industrial and residential activities in the urban economy. In this chapter we study the dynamic property of an urban location model which takes explicit account of the interaction of firms and households in space; particularly, (1) competition between firms and households for urban land, and (2) industrial influence on household income. For this purpose, we consider a model of a circular city which consists of business and residential areas, where production takes place in the business area and housing is provided in the residential area. In order to focus on the dynamic interaction between firms and households, there are assumed to be a single group (industry) of identical firms and a single class of identical households in the city with firms being spatially surrounded by households. It should be noted that the present model is essentially the same as that of Solow (1973) in its static framework, but our emphasis here is placed on dynamics rather than statics.

First, we note that there are two kinds of boundaries to be determined in the model: one between the business and the residential areas, namely, the "CBD boundary," and one between the city and surrounding agricultural land, namely, the "urban boundary." After determining the equilibrium positions of the CBD boundary and the urban boundary, we introduce a dynamic adjustment process of those boundaries in order to examine the stability property of the spatial equilibrium. It is shown to be dynamically stable under the assumptions that (1) firms have a higher ratio of marginal transport cost to land (or at least the same ratio as) households at the CBD boundary, (2) the ratio of residential land to business land at the CBD boundary is no less than the rate of change of the marginal utility of income between the CBD boundary and the urban boundary, and (3) land is a weakly normal good for all households in the sense that the individual demand for land is not decreasing in income.

Next, assuming these stability conditions to be met, we investigate the effect of a change in the opportunity cost of land on the equilibrium boundary positions and the equilibrium wage rate. We can easily show that both the CBD and the urban boundaries will move in as the opportunity cost of land rises and that the wage rate is negatively related to the opportunity cost of land.

3.1   The Urban Location Model

Here we use the notation adopted in the previous chapters, and let C be the unit cost function, V the indirect utility function, u the level of utility, w the household (wage) income, p the product price, N the total number of households (workers), q_b the transport (shipping) cost per unit of product, q_c the transport (commuting) cost per household, r_b the business land rent, r_c the residential land rent, r_0 the opportunity cost of land, x_b the distance from the city center to the CBD boundary, x_c the distance from the city center to the urban boundary, k_b the business land per unit of labor, h_b the business land per unit of output, h_c the residential land per household, g_b a constant fraction of the business area for business use at each distance from the center, and g_c a constant fraction of the residential area for residential use at each distance from the center.

First, the zero-profit condition is expressed as

$$C[w, r_b(x)] = p - q_b(x), \qquad (1)$$

where p is a given constant. Equation (1) can be solved for r_b as

$$r_b(x) = r_b(x, w), \qquad (2)$$

which is the bid rent function of the firms. From Eq. (1), the land–labor ratio becomes

$$k_b(x) = (\partial C/\partial r_b)/(\partial C/\partial w) \equiv k_b[w, r_b(x)]. \qquad (3)$$

If total population N is given as a constant, the full employment condition is then written as

$$2\pi g_b \int_0^{x_b} \frac{x}{k_b[w, r_b(x, w)]}\, dx = N, \tag{4}$$

which gives the wage rate w as a function of x_b, given g_b and N,

$$w = w(x_b). \tag{5}$$

Turning to households, we find

$$V[r_c(x), w - q_c(x)] = u, \tag{6}$$

which can be solved for r_c as

$$r_c(x) = r_c[w - q_c(x), u], \tag{7}$$

which is the bid rent function of the households. Equation (6) also yields

$$h_c(x) = -V_r/V_w \equiv h_c[r_c(x), w - q_c(x)], \tag{8}$$

where $V_r \equiv \partial V/\partial r_c$ and $V_w \equiv \partial V/\partial w$. Since the total number of households is N, the full accommodation condition becomes

$$2\pi g_c \int_{x_b}^{x_c} \frac{x}{h_c\{r_c[w - q_c(x), u], w - q_c(x)\}}\, dx = N, \tag{9}$$

which determines u as a function of w, x_b, and x_c, given g_c and N,

$$u = u(w, x_b, x_c). \tag{10}$$

In the land market, the equilibrium rent function should be continuous at x_b and at x_c. From the bid rent functions (2) and (7), we can express the market rent condition as

$$r_b(x_b, w) = r_c[w - q_c(x_b), u], \qquad r_c[w - q_c(x_c), u] = r_0, \tag{11}$$

where r_0 is a given constant.

In order to see how the equilibrium positions of the boundaries are determined in the system (1)–(11), we substitute Eqs. (5) and (10) into (11) to have

$$r_b[x_b, w(x_b)] = r_c\{w(x_b) - q_c(x_b), u[w(x_b), x_b, x_c]\},$$
$$r_c\{w(x_b) - q_c(x_c), u[w(x_b), x_b, x_c]\} = r_0, \tag{12}$$

which together can be solved for x_b and x_c, given r_0. In order to focus on dynamics, we simply assume the existence of equilibrium values x_b^* and x_c^* in the following analysis.

3.2 Dynamic Stability

Let us investigate the dynamic stability property of the equilibrium. As in the previous chapters, we specify a dynamic adjustment process of the boundaries as

$$\dot{x}_b = f_b(r_b[x_b, w(x_b)] - r_c\{w(x_b) - q_c(x_b), u[w(x_b), x_b, x_c]\}),$$
$$\dot{x}_c = f_c(r_c\{w(x_b) - q_c(x_c), u[w(x_b), x_b, x_c]\} - r_0), \tag{13}$$

where $f_b'(\ ) > 0$, $f_c'(\ ) > 0$, and $f_b(0) = f_c(0) = 0$. Linearizing system (13) around the equilibrium, we find

$$\begin{bmatrix} x_b \div x_b^* \\ x_c - x_c^* \end{bmatrix} = \begin{bmatrix} f_b'(0) & 0 \\ 0 & f_c'(0) \end{bmatrix} \begin{bmatrix} P & Q \\ R & S \end{bmatrix} \begin{bmatrix} x_b - x_b^* \\ x_c - x_c^* \end{bmatrix}, \tag{14}$$

and

$$P \equiv r_{bx}(x_b) + r_{cw}(x_b)q_c'(x_b) + [r_{bw}(x_b) - r_{cw}(x_b)]w_b$$
$$- r_{cu}(x_b)(u_w w_b + u_b), \tag{15}$$

$$Q \equiv -r_{cu}(x_b)u_c, \tag{16}$$

$$R \equiv r_{cw}(x_c)w_b + r_{cu}(x_c)(u_w w_b + u_b), \tag{17}$$

$$S = -r_{cw}(x_c)q_c'(x_c) + r_{cu}(x_c)u_c, \tag{18}$$

with all the variables being evaluated at the equilibrium, where

$$r_{bx}(x) \equiv \partial r_b(x)/\partial x, \qquad r_{bw}(x) \equiv \partial r_b(x)/\partial w,$$
$$r_{cw}(x) \equiv \partial r_c(x)/\partial w, \qquad r_{cu}(x) \equiv \partial r_c(x)/\partial u,$$
$$w_b \equiv dw/dx_b, \qquad u_w \equiv \partial u/\partial w, \qquad u_b \equiv \partial u/\partial x_b, \qquad u_c \equiv \partial u/\partial x_c.$$

It is well known that in the linear system (14) the equilibrium is stable for any positive speeds of adjustment $f_b'(0) > 0$ and $f_c'(0) > 0$, if

$$P < 0, \qquad S < 0, \qquad PS - QR > 0. \tag{19}$$

In this case, stability follows from the fact that the trace and the determinant of the product of the two matrices in system (14) are

$$f_b'(0)P + f_c'(0)S < 0, \qquad f_b'(0)f_c'(0)(PS - QR) > 0, \tag{20}$$

respectively. Stability can also be shown diagrammatically, as seen in Fig. 3.1. It turns out that the stability condition (19) is implied by the following

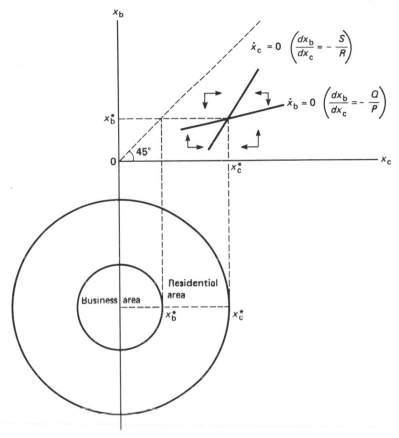

FIGURE 3.1

three conditions together:

$$q_b'(x_b^*)/h_b(x_b^*) \geq q_c'(x_b^*)/h_c(x_b^*), \tag{21}$$

i.e., the ratio of marginal transport cost to land for firms is at least equal to that for households at the CBD boundary;

$$h_c(x_b^*)/h_b(x_b^*) \geq [V_w(x_c^*) - V_w(x_b^*)]/V_w(x_b^*), \tag{22}$$

where $V_w(x) \equiv \partial V(x)/\partial w$, that is to say, the ratio of residential land to business land at the CBD boundary is no less than the rate of change of the marginal utility of income between the CBD boundary and the urban boundary; and

$$\partial h_c[r_c(x), w - q_c(x)]/\partial w \geq 0 \quad \text{for} \quad x_b^* \leq x \leq x_c^*, \tag{23}$$

i.e., residential land is a "weakly" normal good for all households in the sense that the individual demand for land is not decreasing in income.

Theorem 3.1 *In system* (13), *the equilibrium is locally stable for any positive speeds of adjustment* $f'_b(0) > 0$ *and* $f'_c(0) > 0$, *if conditions* (21)–(23) *are met.*

Proof *Step 1*   First, we find $Q > 0$, since $r_{cu} = 1/V_r < 0$ and $u_c > 0$. The latter follows from (9) and $\partial h_c/\partial r_c < 0$ which is due to (23). These properties, with $r_{cw} = -V_w/V_r = 1/h_c > 0$, yield $S < 0$.
Step 2   To show $P < 0$, define

$$K(x) \equiv -r_{cw}(x)/r_{cu}(x), \qquad J \equiv r_{bw}(x_b)/r_{cu}(x_b), \tag{24}$$

where we have $K(x) = (V_w/V_r)/(1/V_r) = V_w(x)$. Then,

$$\partial K(x)/\partial x = \partial V_w(x)/\partial x = (V_{ww} + V_{wr}r_{cw})[-q'_c(x)]$$
$$= (V_{ww}V_r - V_{wr}V_w)[-q'_c(x)/V_r] = \{\partial[1/h_c(x)]/\partial w\}V_r q'_c(x)$$
$$\text{for}\quad x_b \leqq x \leqq x_c, \tag{25}$$

from (23), where we utilize the relation

$$\partial[1/h_c(x)]/\partial w = \partial(-V_w/V_r)/\partial w = -(V_{ww}V_r - V_{rw}V_w)/(V_r)^2.$$

Also noting that $K(x) > 0$ and $J > 0$, we find from (25) that

$$J + H(x_b) \geqq K(x_b) \geqq K(x) \qquad \text{for}\quad x_b \leqq x \leqq x_c,$$

which, together with (24), implies

$$J + K(x_b) \geqq -\frac{\int_{x_b}^{x_c} \frac{1}{(h_c)^2}\left(\frac{\partial h_c}{\partial r_c}r_{cw} + \frac{\partial h_c}{\partial w}\right)dx}{\int_{x_b}^{x_c} \frac{1}{(h_c)^2}\frac{\partial h_c}{\partial r_c}r_{cu}\,dx} = \frac{\partial u}{\partial w}, \tag{26}$$

where the last equality comes from (9). Thus,

$$P = -\frac{q'_b(x_b)}{h_b(x_b)} + \frac{q^i_c(x_b)}{h_c(x_b)} + [J + K(x_b) - u_w]r_{cu}(x_b)w_b - r_{cu}(x_b)u_b < 0,$$

from (21), (26) and

$$r_{cu}(x_b) < 0, \qquad u_b < 0, \qquad w_b > 0, \tag{27}$$

all of which easily follow under the present assumptions. (In fact, we have proved the first two properties in Chapter 2 and the last property in Chapter 1.)

Step 3 We also find

$$
\begin{aligned}
PS - QR &= \{[r_{bx}(x_b) - r_{cw}(x_b)q_c'(x_b)] + [r_{bw}(x_b) - r_{cw}(x_b) \\
&\quad - r_{cu}(x_b)u_w]w_b - r_{cu}(x_b)u_b\}\{-r_{cw}(x_c)q_c'(x_c) + r_{cu}(x_c)u_c\} \\
&\quad - \{-r_{cu}(x_b)u_c\}\{[r_{cw}(x_c) + r_{cu}(x_c)u_w]w_b + r_{cu}(x_c)u_b\} \\
&= [r_{bx}(x_b) - r_{cw}(x_b)q_c'(x_b)][-r_{cw}(x_c)q_c'(x_c) + r_{cu}(x_c)u_c] \\
&\quad + \{[r_{bw}(x_b) - r_{cw}(x_b) - r_{cu}(x_b)u_w]w_b - r_{cu}(x_b)u_b\}\{-r_{cw}(x_c)q_c'(x_c)\} \\
&\quad + \{[r_{hw}(x_b) - r_{cw}(x_b)]r_{cu}(x_c) + r_{cw}(x_c)r_{cu}(x_b)\}w_b u_c \\
&= \left[\frac{q_b'(x_b)}{h_b(x_b)} - \frac{q_c'(x_b)}{h_c(x_b)}\right]C_1 - \{[J + K(x_b) - u_w]r_{cu}(x_b)w_b\}C_2 \\
&\quad + [J + K(x_b) - K(x_c)]C_3 > 0,
\end{aligned}
\tag{28}
$$

where C_1, C_2, and C_3 are all positive.   Q.E.D.

Comparing this theorem with the stability results in the previous chapters, we realize that condition (22) is added to the two familiar conditions, namely, the von Thünen condition and the normality condition, although the latter two conditions in the present version are somewhat weaker than their counterparts in the previous chapters. It is noted that condition (21) concerns the direct effect of a change in the CBD boundary on the rent difference at the CBD boundary, holding the wage rate constant, whereas condition (22) ensures that the indirect effect of the boundary change through a change in the wage rate is in the same direction as the direct effect.

In order to gain more insight into these stability conditions, let us assume the production function and the utility function to be Cobb–Douglas, so that we can write

$$
C = (r_b)^{auw^{1-a}}, \qquad V = (r_c)^{-b}[w - q_c(x)].
\tag{29}
$$

In this case, the normality condition (23) is automatically satisfied, and the von Thünen condition (21) becomes

$$
q_c'(x_b)/q_b'(x_b) \leqq b[w - q_c(x_b)]/a[p - q_b(x_b)].
\tag{30}
$$

Since $V_w = u/[w - q_c(x)]$ from (29), condition (22) can be written as

$$
[q_c(x_c) - q_c(x_b)]/[w - q_c(x_c)] \leqq b[w - q_c(x_b)]/a[p - q_b(x_b)].
\tag{31}
$$

In fact, both conditions (30) and (31) can be met, if marginal commuting cost $q_c'(x)$ is sufficiently small everywhere in the residential area, and particularly small relative to marginal shipping cost $q_b'(x)$ at the CBD boundary.

3.3   Comparative Statics

We are now ready to carry out a comparative static analysis to find the effect of a change in the opportunity cost of land on the equilibrium boundary positions and the equilibrium wage rate. Applying the Correspondence Principle (see Samuelson, 1947), we assume the stability conditions (21)–(23) to be satisfied, and differentiate system (12) with respect to r_0 to obtain

$$\begin{bmatrix} P & Q \\ R & S \end{bmatrix} \begin{bmatrix} dx_b^*/dr_0 \\ dx_c^*/dr_0 \end{bmatrix} = \begin{bmatrix} 0 \\ 1 \end{bmatrix}, \tag{32}$$

which can be solved for

$$dx_b^*/dr_0 = -Q/(PS - RQ), \qquad dx_c^*/dr_0 = P/(PS - RQ), \tag{33}$$

where P, Q, R, and S are defined as in Eqs. (15)–(18). In the proof of Theorem 3.1, we have shown that $Q > 0$, $P < 0$, and $PS - RQ > 0$. Thus, we have established the following proposition.

Theorem 3.2   *If the stability conditions (21)–(23) are met, then*

$$dx_b^*/dr_0 < 0, \qquad dx_c^*/dr_0 < 0, \tag{34}$$

i.e., both the CBD boundary and the urban boundary will move inward, as the opportunity cost of land increases.

The effect of an increase in the opportunity cost of land is illustrated in Fig. 3.2.

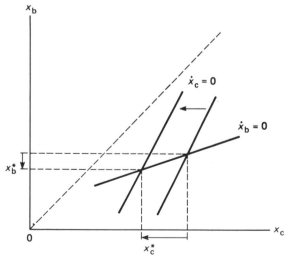

FIGURE 3.2

Finally, it is easy to determine the effect of a change in r_0 on the equilibrium wage rate w. Since w is positively related to x_b from Eq. (27), it follows from Theorem 3.2 that

$$dw^*/dr_0 = (dw^*/dx_b)(dx_b^*/dr_0) < 0. \tag{35}$$

Theorem 3.3   *Under conditions* (21)–(23), *we find*

$$dw^*/dr_0 < 0, \tag{36}$$

i.e., the wage rate will fall, as the opportunity cost of land rises.

REFERENCES

Samuelson, P. A. (1947). *Foundations of Economic Analysis.* Harvard Univ. Press, Cambridge, Massachusetts.
Solow, R. M. (1973). On equilibrium models of urban location. In *Essays in Modern Economics* (M. Parkin, ed.), pp. 2–16. Longman Group, London.

CHAPTER 4

An Open City Case

In the literature, two kinds of residential city models have been developed, namely, "closed" city models and "open" city models.[1] Closed city models are ones in which household class sizes are given exogenously with utility levels to be determined endogenously. These models are called closed because there is assumed to be no migration into or out of the city. On the other hand, in open city models household class sizes are determined endogenously, whereas utility levels are given from outside. In this case, the city is open in the sense that people move freely into and out of the city so as to give each household a certain utility level which can be achieved outside the city. While we have dealt with closed city models in the previous chapters, we have yet to explore the dynamics and comparative statics of an open city.

In this chapter we are mainly concerned with the dynamic stability property of an open city with many classes of households.[2] In fact, the city is assumed to be closed in the short run and open in the long run; i.e., at each moment of time, household class sizes are given with utility levels to be determined within the city, but in the long run, household class sizes are eventually adjusted so that each household will attain a given utility level which is obtainable elsewhere outside the city. We specify a dynamic adjustment process of household movement into and out of the city, depending on the utility levels which can be obtained in the city at each moment of

---

[1] See, e.g., Wheaton (1974) on both closed and open city models.
[2] This chapter is an extended version of Miyao (1979).

time and those attainable outside the city. In equilibrium, there is no household movement into or out of the city.

We prove that the equilibrium is dynamically stable for any positive speeds of adjustment under the same conditions assumed in Chapter 2, namely, the von Thünen condition and the non-Giffen good condition. Then some comparative static properties of the model are derived from the stability analysis. In particular, for each class the total number of households is shown to be inversely related to the level of utility, regardless of whichever is given exogenously. It is also found that the effects of changes in such parameters as household income and the opportunity cost of land in the open city case are quite different from those in the closed city case.

4.1 The Open City Model

As in Chapter 2, there are assumed to be m classes of households with identical income levels, utility functions, and transport cost functions within each class, but possibly different among classes. It is also assumed that the utility of each household depends on the amount of a consumption good and residential space and that the price of the consumption good is given and normalized as unity. Every resident commutes to the CBD, and transport cost is a sole function of distance.

Then, we can derive the maximized level of utility for a household in class i living at distance x from the CBD as a function of land rent and net income,

$$V_i[r_i(x), w_i - q_i(x)] = u_i \qquad (i = 1, \ldots, m), \qquad (1)$$

where V_i is class i's indirect utility function depending on bid rent $x_i(x)$ and net income, i.e., gross income w_i minus transport cost $q_i(x)$, and u_i is the utility level which is common for all households in class i in equilibrium. In view of the well-known properties of the indirect utility function, it follows from Eq. (1) that

$$r_i(x) = r_i[w - q_i(x), u_i] \qquad (i = 1, \ldots, m) \qquad (2)$$

with

$$r_{iw} \equiv \partial r_i/\partial w_i = -(\partial V_i/\partial w_i)/(\partial V_i/\partial r_i) = 1/h_i > 0,$$
$$r_{iu} \equiv \partial r_i/\partial u_i = 1/(\partial V_i/\partial r_i) < 0, \qquad (3)$$

where h_i is residential land per household in class i.

As in Chapter 2, we may express the full accommodation condition that all households in class i be located in zone i as

$$2\pi \int_{x_{i-1}}^{x_i} \frac{g(x)x}{h_i[r_i(x), w_i - q_i(x)]} dx = N_i \qquad (i = 1, \ldots, m), \qquad (4)$$

and the market rent condition that the overall rent function be continuous at every boundary as

$$r_i[w_i - q_i(x_i), u_i] = r_{i+1}[w_{i+1} - q_{i+1}(x_i), u_{i+1}] \qquad (i = 1, \ldots, m-1),$$
$$r_m[w_m - q_m(x_m), u_m] = r_0, \tag{5}$$

in view of Eq. (2).

First, we solve system (5) for x's as functions of u_i's, given w_i's and r_0,

$$x_i = x_i(u_i, u_{i+1}; w_i, w_{i+1}) \qquad (i = 1, \ldots, m-1),$$
$$x_m = x_m(u_m; w_m, r_0). \tag{6}$$

We prove the following lemma.

Lemma 4.1   *In system (6), we find*

$$\partial x_i/\partial u_i < 0, \qquad \partial x_i/\partial u_{i+1} > 0, \qquad \partial x_i/\partial w_i > 0, \qquad \partial x_i/\partial w_{i+1} < 0$$
$$(i = 1, \ldots, m-1),$$

and

$$\partial x_m/\partial u_m < 0, \qquad \partial x_m/\partial w_m > 0, \qquad \partial x_m/\partial r_0 < 0 \tag{7}$$

if the following condition is satisfied:

$$q_i'(x_i)/h_i(x_i) > q_{i+1}'(x_i)/h_{i+1}(x_i) \qquad (i = 1, \ldots, m-1). \tag{8}$$

Proof   Let

$$Q_i \equiv -r_{iw}(x_i)q_i'(x_i) + r_{i+1w}(x_i)q_{i+1}'(x_i) \qquad (i = 1, \ldots, m-1),$$
$$Q_m = -r_{mw}(x_m)q_m'(x_m).$$

It follows from (3) and (8) that

$$Q_i = -[q_i'(x_i)/h_i(x_i) - q_{i+1}'(x_i)/h_{i+1}(x_i)] < 0 \qquad (i = 1, \ldots, m-1)$$
$$Q_m = -q_m'(x_m)/h_m(x_m) < 0. \tag{9}$$

Then, we find from (5) that

$$\partial x_i/\partial u_i = -r_{iu}/Q_i < 0, \qquad \partial x_i/\partial w_i = -r_{iw}/Q_i > 0,$$
$$(i = 1, \ldots, m),$$
$$\partial x_i/\partial u_{i+1} = r_{i+1u}/Q_i > 0, \qquad \partial x_i/\partial w_{i+1} = r_{i+1w}/Q_i < 0,$$
$$(i = 1, \ldots, m-1),$$
$$\partial x_m/\partial r_0 = 1/Q_m < 0. \qquad \text{Q.E.D.}$$

By using Eqs. (2) and (6), we can express the left-hand side of Eq. (4) as a function of u_i's, given w_i's and r_0,

$$M_i \equiv \int_{x_{i-1}(u_{i-1}, u_i; w_{i-1}, w_i)}^{x_i(u_i, u_{i+1}; w_i, w_{i+1})} \frac{s(x)}{h_i\{r_i[w_i - q_i(x), u_i], w_i - q_i(x)\}} \, dx,$$

$$(i = 1, \ldots, m-1), \quad (10)$$

$$M_m \equiv \int_{x_{m-1}(u_{m-1}, u_m; w_{m-1}, w_m)}^{x_m(u_m; w_m, r_0)} \frac{s(x)}{h_m\{r_m[w_m - q_m(x), u_m], w_m - q_m(x)\}} \, dx, \quad (11)$$

where $s(x) \equiv 2\pi g(x)x$. Then, condition (4) can be written as

$$M_i(u_{i-1}, u_i, u_{i+1}) = N_i, \quad (i = 1, \ldots, m), \quad (12)$$

where it is understood that $M_1(\ ) = M_1(u_1, u_2)$ and $M_m = M_m(u_{m-1}, u_m)$.

At each moment of time, the city is closed in the sense that the sizes of household classes N_i are given and the corresponding utility levels u_i are determined endogenously by solving system (12) for u_i as

$$u_i = u_i(N_1, \ldots, N_m) \quad (i = 1, \ldots, m). \quad (13)$$

In the long run, however, the city is open in the sense that the number of households in each class will adjust perfectly so as to equate the utility level u_i obtained in the city to an exogenously given utility level u_i which can be attained elsewhere. From system (12), the long-run equilibrium values of N_i's are found to be

$$N_i^* = M_i(\bar{u}_{i-1}, u_i, u_{i+1}) \quad (i = 1, \ldots, m). \quad (14)$$

In what follows, we assume the existence of a long-run equilibrium with $N_i^* > 0$ $(i = 1, \ldots, m)$. Note that this assumption does not preclude the possibility that certain household classes will be excluded from the city in the long run because of higher utility levels available for those classes elsewhere. In that case, we need only renumber household classes in such a way that we have $N_i^* > 0$ $(i = 1, \ldots, m)$ for those classes which are included in the city, and $N_j^* = 0$ $(j = m + 1, \ldots)$ for those which are excluded in the long run.

4.2 Dynamic Adjustment and Stability

We now introduce a dynamic adjustment process of household movement into and out of the city. The number of households in each class is assumed to increase *or* decrease through time, due to in-migration or out-migration, as the utility level those households can attain in the city is higher *or* lower than a certain level which is given from outside. More specifically, we assume

$$\dot{N}_i = f_i[u_i(N_1,\ldots,N_m) - \bar{u}_i] \qquad (i = 1,\ldots,m) \qquad (15)$$

with $f_i'(\ ) > 0$, $f_i(0) = 0$, where the dot denotes differentiation with respect to time. From Eqs. (12)–(14), it is clear that we have $\dot{N}_i = 0$ $(i = 1,\ldots,m)$ in equilibrium.

Now we prove the following theorem.

Theorem 4.1   *According to process* (15), *the equilibrium is locally stable for any positive speeds of adjustment $f_i'(0) > 0$ $(i = 1,\ldots,m)$ if the von Thünen condition*

$$q_i'(x_i^*)/h_i(x_i^*) > q_{i+1}'(x_i^*)/h_{i+1}(x_i^*) \qquad (i = 1,\ldots,m-1) \qquad (16)$$

and the non-Giffen good condition

$$\partial h_i[r_i(x), w_i - q_i(x)]/\partial r_i < 0 \qquad \text{for} \quad x_{i-1}^* \leqq x \leqq x_i^*$$
$$(i = 1,\ldots,m) \qquad (17)$$

are satisfied.

Proof   Total differentiation of (12) with respect to u_i and N_i $(i = 1,\ldots,m)$ gives $A\,du = dN$, where the ith element of du is du_i and the ith element of dN is dN_i $(i = 1,\ldots,m)$; and the matrix A is such that

$$a_{ii-1} \equiv -[s(x_{i-1})/h_i(x_{i-1})]\,\partial x_{i-1}/\partial u_{i-1} \qquad (i = 2,\ldots,m),$$

$$a_{ii} \equiv \int_{x_{i-1}}^{x_i} \frac{\partial(1/h_i)}{\partial r_i} r_{iu} s(x)\,dx - \frac{s(x_{i-1})}{h_i(x_{i-1})}\frac{\partial x_{i-1}}{\partial u_i} + \frac{s(x_i)}{h_i(x_i)}\frac{\partial x_i}{\partial u_i}$$

$$(i = 1,\ldots,m),$$

$$a_{ii+1} \equiv \frac{s(x_i)}{h_i(x_i)}\frac{\partial x_i}{\partial u_{i+1}} \qquad (i = 1,\ldots,m-1), \qquad (18)$$

and all other elements are zero.[3] In view of (3), (7), and (16)–(18), we can show that the matrix A, evaluated at the equilibrium, has negative diagonal elements and is quasi-dominant diagonal, i.e., there exist positive constants

$$c_1 \equiv 1, \qquad c_i \equiv \prod_{k=2}^{i}\left(-\frac{\partial x_{i-1}/\partial u_{i-1}}{\partial x_{i-1}/\partial u_i}\right) \qquad (i = 2,\ldots,m) \qquad (19)$$

such that

$$c_1|a_{11}| > c_2|a_{12}|,$$

$$c_i|a_{ii}| > c_{i-1}|a_{ii-1}| + c_{i+1}|a_{ii+1}| \qquad (i = 2,\ldots,m-1),$$

$$c_m|a_{mm}| > c_{m-1}|a_{mm-1}|.$$

---

[3] It is worth noting that we find qualitatively the same matrix in the stability analysis of a closed city in Chapter 2.

Thus, according to McKenzie (1960), A is nonsingular and every characteristic root of A has a negative real part.[4] On the other hand, there exist positive diagonal matrices C and B with their ith diagonal elements c_i as defined in (19) and

$$b_1 \equiv 1, \qquad b_i \equiv \prod_{k=2}^{i} \left[\frac{h_i(x_{i-1})}{h_{i-1}(x_{i-1})} \right] \qquad (i = 2, \ldots, m), \qquad (20)$$

respectively, such that $BAC = S$, where S is a symmetric matrix. Then we can find a diagonal matrix D and a nonsingular matrix Z such that $BACZ = SZ = DZ$, where the diagonal elements of D correspond to m real characteristic roots of S. Since B and C are both nonsingular, it follows that $ACZ = B^{-1}DC^{-1}CZ$, which means that the characteristic roots of A are the diagonal elements of $B^{-1}DC^{-1}$ and therefore are all real. Thus, all the roots of A are real and negative, and so are all the roots of A^{-1}. Finally, by linearizing (15) around the equilibrium, we find

$$(N \dot{-} N^*) = FA^{-1}(N - N^*),$$

where the ith element of $(N - N^*)$ is $N_i - N_i^*$, and F is a diagonal matrix with its ith element $f_i'(0) > 0$ $(i = 1, \ldots, m)$. Since all the roots of A^{-1} are negative, the theorem has been proved.   Q.E.D.

4.3   Comparative Statics

Wheaton (1974) has shown that in the case of one homogeneous household class, the level of utility u is negatively related to class size N in equilibrium. More specifically, in the closed city with N given, we find $du/dN < 0$, whereas in the open city with u given, we have $dN/du < 0$. Hartwick et al. (1976) have generalized this result in the closed city case with many household classes under the assumption that land is a normal good for all classes. Now we can relax the normality assumption and give a much simpler proof than theirs by making use of our stability analysis.

Theorem 4.2   *In the closed city with N_i exogenous and u_i endogenous $(i = 1, \ldots, m)$, we have*

$$du_i/dN_j < 0 \qquad \text{for all} \quad i, j = 1, \ldots, m, \qquad (21)$$

if the von Thünen condition (16) and the non-Giffen good condition (17) are satisfied.

---

[4] Since A is nonsingular, we may solve (12) for u_i in equilibrium so that the differential equation system (15) makes sense in the small neighborhood of equilibrium.

Proof As seen in the proof of Theorem 4.1, we find $A\, du = dN$, and A is indecomposable and quasi-dominant diagonal with negative diagonal elements and nonnegative off-diagonal elements. Thus, $du = A^{-1}\, dN$, where all the elements of A^{-1} are negative (see Quirk and Saposnik, 1968, p. 212). Q.E.D.

We can now generalize Wheaton's result in the open city case with many household classes.

Theorem 4.3 *In the open city with $\bar{u}_i$ exogenous and N_i^* endogenous* $(i = 1, \ldots, m)$, *we obtain*

$$dN_i^*/d\bar{u}_i < 0, \qquad dN_i^*/d\bar{u}_{i-1} > 0, \qquad dN_i^*/d\bar{u}_{i+1} > 0,$$

$$dN_i^*/d\bar{u}_j = 0 \qquad for\ all \quad j \neq i - 1, i, i + 1 \qquad (i = 1, \ldots, m),$$

(22)

if conditions (16) *and* (17) *are met.*

Proof In the proof of Theorem 4.1, we find $a_{ii} < 0$, $a_{ii-1} > 0$, $a_{ii+1} > 0$, and $a_{ij} = 0$, for all $j \neq i - 1, i, i + 1$ $(i = 1, \ldots, m)$. Q.E.D.

In this case, class i's utility level u_i will affect only the size of its own and its immediate neighbors, namely, classes $i - 1$ and $i + 1$, by shifting class i's bid rent curve alone, with the bid rent curves of all the other classes unaffected, as illustrated in Fig. 4.1. Therefore, the following property is intuitively obvious and can be easily derived from Lemma 4.1.

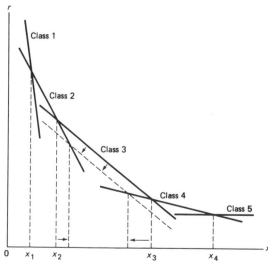

FIGURE 4.1

Theorem 4.4   *In the open city with $\bar{u}_i$ exogenous $(i = 1,\ldots,m)$, we have*

$$dx_i^*/d\bar{u}_i < 0, \qquad dx_i^*/d\bar{u}_{i+1} > 0,$$
$$dx_i^*/d\bar{u}_j = 0 \qquad \text{for all } j \neq i, i+1 \qquad (i = 1,\ldots,m),$$

(23)

if the von Thünen condition (16) is satisfied.

Note that this theorem holds without even assuming that land is a non-Giffen good.

As for the effect of a change in household income, we can show the following theorem.

Theorem 4.5   *In the open city with $\bar{u}_i$ given exogenously $(i = 1,\ldots,m)$, we find*

$$dx_i^*/dw_i > 0, \qquad dx_i^*/dw_{i+1} < 0,$$
$$dx_i^*/dw_j = 0 \qquad \text{for all } j \neq i, i+1, \qquad (i = 1,\ldots,m),$$

(24)

and

$$dN_i^*/dw_{i-1} < 0, \qquad dN_i^*/dw_i > 0, \qquad dN_i^*/dw_{i+1} < 0,$$
$$dN_i^*/dw_j = 0 \qquad \text{for all } j \neq i-1, i, i+1 \qquad (i = 1,\ldots,m),$$

(25)

if condition (16) is met.

Proof   (24) follows directly from (6) and Lemma 4.1. To prove (25), we totally differentiate (12) with respect to w_i and N_i and find $\bar{A}\,dw = dN$, where the ith element of dw is dw_i, and the matrix $\bar{A}$ is such that

$$\bar{a}_{ii-1} \equiv \frac{\partial N_i}{\partial w_{i-1}} = -\frac{s(x_{i-1})}{h_i(x_{i-1})}\frac{\partial x_{i-1}}{\partial u_{i-1}} < 0 \qquad (i = 2,\ldots,m),$$

$$\bar{a}_{ii} \equiv \frac{\partial N_i}{\partial w_i} = \int_{x_{i-1}}^{x_i}\left[\frac{\partial(1/h_i)}{\partial r_i}r_{iw} + \frac{\partial(1/h_i)}{\partial w_i}\right]s(x)\,dx - \frac{s(x_{i-1})}{h_i(x_{i-1})}\frac{\partial x_{i-1}}{\partial w_i}$$
$$+ \frac{s(x_i)}{h_i(x_i)}\frac{\partial x_i}{\partial w_i} > 0 \qquad (i = 1,\ldots,m),$$

where the expression inside the integral is always positive, because $1/h_i$ is increasing in r_i with a "compensating" change in w_i to maintain the utility level $\bar{u}_i$,[5] and

$$\bar{a}_{ii+1} \equiv \frac{\partial N_i}{\partial w_{i+1}} = \frac{s(x_i)}{h_i(x_i)}\frac{\partial x_i}{\partial u_{i+1}} < 0 \qquad (i = 1,\ldots,m-1).$$

Obviously, all the other elements of $\bar{A}$ are zero.   Q.E.D.

---

[5] In other words, this is due to the negative substitution effect.

Finally, it is immediate from Lemma 4.1 that a change in the opportunity cost of land r_0 will affect only the outer city boundary x_m while leaving all the other boundaries x_i $(i = 1, \ldots, m - 1)$ unchanged.

Theorem 4.6 *In the open city with $\bar{u}_i$ exogenous $(i = 1, \ldots, m)$, we find*

$$dx_m^*/dr_0 < 0, \qquad dx_i^*/dr_0 = 0 \qquad \text{for all} \quad i = 1, \ldots, m - 1, \qquad (26)$$

if the von Thünen condition (16) is satisfied.

Comparing Theorems 4.5 and 4.6 with the comparative static results for the closed city in the previous chapters, we conclude that the effects of changes in such parameters as household income and the opportunity cost of land are rather limited, or "localized," in the open city case as compared to the closed city, in which case those effects are felt more or less everywhere in the city. This difference is due to the fact that in the open city case with utility levels given, the positions of all the bid rent curves are independent of each other and, therefore, a shift in a bid rent curve does not affect the position of any other curve, whereas in the closed city case, all the bid rent curves are mutually dependent and affect each other through endogenous changes in utility levels.[6]

REFERENCES

Hartwick, J., Schweizer, U., and Varaiya, P. (1976). Comparative statics of a residential economy with several classes. *Journal of Economic Theory* **13**, 396–413.

McKenzie, L. W. (1960). Matrices with dominant diagonals and economic theory. In *Proceedings of a Symposium on Mathematical Methods in the Social Science*, pp. 277–292. Stanford Univ. Press, Stanford, California.

Miyao, T. (1979). Dynamic stability of an open city with many household classes. *Journal of Urban Economics* **6**, 292–298.

Polinsky, A. M., and Shavell, S. (1976). Amenities and property values in a model of an urban area. *Journal of Public Economics* **5**, 119–129.

Quirk, J., and Saposnik, R. (1968). *Introduction to General Equilibrium Theory and Welfare Economics*. McGraw-Hill, New York.

Wheaton, W. C. (1974). A comparative static analysis of urban spatial structure. *Journal of Economic Theory* **9**, 223–237.

[6] For somewhat related results regarding open and closed cities, see Polinsky and Shavell (1976).

PART **2**

EXTERNALITY AND INSTABILITY

CHAPTER 5

Instability of a Mixed City

In this chapter we introduce some elements of externalities into our urban location model and investigate how the dynamic property of the system is affected by the presence of externalities.[1] In particular, we consider the kind of externalities which can arise from the interactions of various classes or groups of households within a city. Our result shows that the system tends to lose its stable nature if certain types of neighborhood externalities are present.

Actually, it is often asked whether a city, which currently accommodates a variety of economic, social, and racial groups, will remain a mixed city or become exclusively occupied by a single group, e.g., poor blacks, with all other groups of households moving out of the city in the long run. In order to answer this question properly, it seems essential to develop a long-run dynamic model of a city with emphasis on socioeconomic factors such as neighborhood externalities within or among various groups of households as well as spatial factors such as the individual choice of residential space and location. In the literature, the most often cited work which employs a long-run dynamic approach to this kind of problem is that of Schelling (1969, 1971). Although Schelling's socioeconomic analysis has revealed some of the dynamic properties of a racially mixed area with neighborhood externalities, his model itself is not quite satisfactory from the economic point of view, since it lacks individual utility functions and thereby disregards the aspect of the individual choice of space and location within the city. As a result,

---

[1] This chapter is based on Miyao (1978).

he failed to analyze the effect of spatial segregation within the city on the long-run nature of the residential composition of the city.[2]

In this chapter we develop a long-run dynamic model of an open city which accommodates two groups of households (e.g., blacks and whites, rich and poor, young and old) with their utility functions depending on a consumption good and residential space. We also assume two kinds of neighborhood externalities. The first is called "negative *inter*group externalities"; i.e., the utility of a typical household in one group is adversely affected by an increase in the number of households in another group living in the city. The second kind is called "positive *intra*group externalities"; i.e., the utility of a typical household in one group is positively affected by an increase in the size of its own group in the city.

We define a "mixed-city" equilibrium as an equilibrium which allows the two groups to coexist in the city in the long run, and then introduce a dynamic adjustment process of household movements into and out of the city. It is proved that the mixed-city equilibrium is dynamically unstable in the presence of negative intergroup externalities and/or positive intragroup externalities, if residential land is perfectly homogeneous and there is no differential transport cost incurred within the city. It turns out, however, that in the case of a monocentric city with positive transport cost, as assumed in the previous chapters, the mixed-city equilibrium may be stable or unstable depending on the degree of neighborhood externalities, because of the stabilizing effect of spatial segregation between the two groups within the city.

5.1 Negative Intergroup Externalities

There are assumed to be two groups of households, e.g., blacks and whites, living together in a city with N_1 households in group 1 and N_2 households in group 2. In the city the total amount of residential land is given and fixed exogenously. In this section and the following two sections, we suppose that land is perfectly homogenous in quality, and transportation cost (more

[2] Although spatial aspects are taken account of in some of the recent work related to the present subject, those models are all completely static. For instance, see Rose-Ackerman (1975), Yinger (1976), and White (1977). Quite recently, Anas (1980) has developed a dynamic model of residential change to explain the phenomenon of neighborhood tipping without assuming intergroup or intragroup externalities. Since our interest is in the effect of such externalities, his work and ours are complementary.

rigorously, differential transport cost) is negligible in the city, so that there is no economic reason for a household to prefer one location to another within the city. Note that this assumption is different from the monocentric city assumption made in the previous chapters.

Here we introduce some negative intergroup externalities, i.e., each group does not like the other. More specifically, the utility of a typical household in one group tends to decrease with an increase in the number of households in the other group living in the city. This means that the utility U_i of a typical household in group i depends not only on the amount of a consumption good d_i and the amount of residential land (space) h_i, but also on the size of the other group N_j in the city,

$$U_i = U_i(d_i, h_i, N_j), \qquad i \neq j \qquad (i, j = 1, 2), \tag{1}$$

where negative intergroup externalities are expressed as

$$\partial U_i / \partial N_j < 0, \qquad i \neq j \qquad (i, j = 1, 2). \tag{2}$$

As a first approximation, it seems reasonable to assume that for given d_i and h_i, neighborhood externalities will affect the level of utility without changing the marginal rate of substitution between d_i and h_i, so that the utility function can be written in the separable form

$$U_i = F_i(d_i, h_i) E_i(N_j), \qquad i \neq j \qquad (i, j = 1, 2), \tag{3}$$

where

$$E_i'(N_j) < 0, \qquad i \neq j \qquad (i, j = 1, 2). \tag{4}$$

and the function F_i is assumed to possess all the desirable properties such as twice-differentiability and concavity as usually assumed in the literature.

Given a certain amount of income $w_i > 0$, each household maximizes its utility (3) subject to the budget constraint $d_i + rh_i = w_i$ $(i = 1, 2)$, where r is land rent and the price of the consumption good is given and normalized as unity.[3] Then, the corresponding indirect utility function V_i takes the form

$$V_i = G_i(r, w_i) E_i(N_j), \qquad i \neq j \qquad (i, j = 1, 2), \tag{5}$$

from which it follows that

$$\partial V_i / \partial r = (\partial G_i / \partial r) E_i(N_j) < 0, \qquad \partial V_i / \partial w_i = (\partial G_i / \partial w_i) E_i(N_j) > 0, \tag{6}$$

and

$$h_i = -(\partial V_i / \partial r)/(\partial V_i / \partial w_i) = -(\partial G_i / \partial r)/(\partial G_i / \partial w_i) \equiv h_i(r, w_i). \tag{7}$$

---

[3] As in the previous chapters, we are here dealing with a small city relative to the national economy so that the price of the consumption good is given exogenously in this city.

With the total amount of residential land L given exogenously, we can equate total supply and total demand for land as

$$h_1(r, w_1)N_1 + h_2(r, w_2)N_2 = L. \tag{8}$$

If we assume that for any given $w_i > 0$,

$$\partial h_i(r, w_i)/\partial r < 0 \qquad (i = 1, 2), \tag{9}$$

i.e., residential land is a non-Giffen good, and furthermore[4]

$$h_i(0, w_i) = \infty, \qquad h_i(\infty, w_i) = 0 \qquad (i = 1, 2),$$

then r is uniquely determined by Eq. (8) as a function of N_1 and N_2, given w_1, w_2, and L,

$$r = r(N_1, N_2) \tag{10}$$

with

$$\partial r/\partial N_i = -h_i/\left[(\partial h_1/\partial r)N_1 + (\partial h_2/\partial r)N_2\right] > 0 \qquad (i = 1, 2), \tag{11}$$

for $N_1 \geqq 0$ and $N_2 \geqq 0$, but not $N_1 = N_2 = 0$.

At each moment of time, the city is "closed" in the sense that group sizes N_1 and N_2 are given, whereas utility levels u_1 and u_2 are determined from Eqs. (5) and (10) as

$$u_i = u_i(N_1, N_2) = G_i\left[r(N_1, N_2), w_i\right]E_i(N_j), \qquad i = j \qquad (i, j = 1, 2). \tag{12}$$

In the long run, however, the city is "open" in the sense that group sizes will be completely adjusted through in- and out-migration so as to equate the utility level u_i of the typical household in group i with a certain utility level $\bar{u}_i$ which is attainable elsewhere outside the city. More precisely, a long-run equilibrium is characterized by $N_1^* \geqq 0$ and $N_2^* \geqq 0$ such that

$$u_i(N_i^*, N_2^*) \leqq \bar{u}_i, \qquad \left[u_i(N_1^*, N_2^*) - \bar{u}_i\right]N_i^* = 0 \qquad (i = 1, 2), \tag{13}$$

where u_1 and u_2 are given exogenously. Obviously, $N_i^* = 0$ if $u_i < \bar{u}_i$ in equilibrium, which means that no one in group i will stay in the city because he will be better off elsewhere.

In order to examine the stability property of an equilibrium, we introduce a dynamic adjustment process of household movement into and out of the city, just as in the previous chapter. Group size N_i is assumed to be increasing or decreasing gradually through time, due to in-migration or out-migration,

---

[4] These two conditions, along with the non-Giffen good condition, are all satisfied in the case of Cobb–Douglas utility functions which will be assumed later.

as the utility level $u_i(N_1, N_2)$ is higher or lower than $\bar{u}_i$ at each moment of time,

$$\dot{N}_i = \begin{cases} f_i[u_i(N_1, N_2) - \bar{u}_i], & \text{when } N_i > 0, \\ f_i[\max\{0, [u_i(N_1, N_2) - \bar{u}_i]\}], & \text{when } N_i = 0 \quad (i = 1, 2), \end{cases} \quad (14)$$

with $f_i'(\ ) > 0$ and $f_i(0) = 0$, where the dot denotes differentiation with respect to time. It is clear that the equilibrium condition (13) is equivalent to $\dot{N}_i = 0 \ (i = 1, 2)$.

Let us define a mixed-city equilibrium as $N_1^* > 0$ and $N_2^* > 0$ satisfying condition (13), i.e.,

$$u_i(N_1^*, N_2^*) = \bar{u}_i \qquad (i = 1, 2). \quad (15)$$

Assume the existence of a mixed-city equilibrium or, in other words, assume that u_1 and u_2 are given in such a way that a mixed-city equilibrium exists. Then, we can prove the following theorem.

Theorem 5.1   *According to process* (14), *the mixed-city equilibrium is locally unstable in the presence of negative intergroup externalities, if residential land is a non-Giffen good for every household.*

Proof   By linearizing (14) in a small neighborhood of the equilibrium with $N_1^* > 0$ and $N_2^* > 0$, we find

$$\begin{bmatrix} \dot{N}_1 - N_1^* \\ \dot{N}_2 - N_2^* \end{bmatrix} = \begin{bmatrix} a_{11} & a_{12} \\ a_{21} & a_{22} \end{bmatrix} \begin{bmatrix} N_1 - N_1^* \\ N_2 - N_2^* \end{bmatrix}, \quad (16)$$

with

$$a_{11} \equiv \partial f_1(0)/\partial N_1 = f_1'(0)(\partial G_1/\partial r)(\partial r/\partial N_1)E_1 < 0,$$
$$a_{12} \equiv \partial f_1(0)/\partial N_2 = f_1'(0)[(\partial G_1/\partial r)(\partial r/\partial N_2)E_1 + G_1 E_1'(N_2)] < 0,$$
$$a_{21} \equiv \partial f_2(0)/\partial N_1 = f_2'(0)[(\partial G_2/\partial r)(\partial r/\partial N_1)E_2 + G_2 E_2'(N_1)] < 0, \quad (17)$$
$$a_{22} \equiv \partial f_2(0)/\partial N_2 = f_2'(0)(\partial G_2/\partial r)(\partial r/\partial N_2)E_2 < 0,$$

where all the variables are evaluated at the equilibrium. It is easy to see from (17) together with (4), (6), and (11) that

$$\begin{aligned} a_{11}a_{22} - a_{12}a_{21} = -f_1'(0)f_2'(0)\big[& G_1 E_1'(N_2)(\partial G_2/\partial r)(\partial r/\partial N_1)E_2 \\ & + G_2 E_2'(N_1)(\partial G_1/\partial r)(\partial r/\partial N_2)E_1 \\ & + G_1 E_1'(N_2)G_2 E_2'(N_1)\big] < 0. \end{aligned}$$

As is well known, a necessary and sufficient condition for stability is that $a_{11} + a_{22} < 0$ and $a_{11}a_{22} - a_{12}a_{21} > 0$, the latter being violated in the present case. Q.E.D.

5.2 Global Analysis in a Special Case

In this section we proceed further to examine the global stability property of system (14) by assuming a Cobb–Douglas utility function,

$$U_i = (d_i)^{a_i}(h_i)^{b_i}(C_i + N_j)^{-c_i}, \qquad i = j \qquad (i, j = 1, 2), \tag{18}$$

where a_i, b_i, c_i, and C_i ($i = 1, 2$) are all positive constants. Note that the assumption $C_i > 0$ yields a finite utility level for group i even when no household in group j is living in the city.

In this special case, the indirect utility function will be

$$V_i = A_i r^{-b_i}(w_i)^{a_i + b_i}(C_i + N_j)^{-c_i}, \qquad i \neq j \quad (i, j = 1, 2), \tag{19}$$

where

$$A_i \equiv (a_i)^{a_i}(b_i)^{b_i}(a_i + b_i)^{-(a_i + b_i)} > 0 \qquad (i = 1, 2).$$

Then we derive

$$h_i = \beta_i w_i / r \qquad (i = 1, 2), \tag{20}$$

where $\beta_i \equiv b_i/(a_i + b_i)$, which is the constant proportion of rent payment to income. From Eqs. (8) and (20) we can obtain r as

$$r = (\beta_1 w_1 N_1 + \beta_2 w_2 N_2)/L. \tag{21}$$

Then, Eqs. (19) and (21) give

$$u_i(N_1, N_2) = \alpha_i(\beta_1 w_1 N_1 + \beta_2 w_2 N_2)^{-b_i}(C_i + N_j)^{-c_i},$$
$$i = j \quad (i, j = 1, 2), \tag{22}$$

where $\alpha_i \equiv A_i L^{b_i}(w_i)^{a_i + b_i}$, which is a given constant.

By setting $u_1(\ ) = \bar{u}_1$ and $u_2(\ ) = \bar{u}_2$ alternately, we can find

$$0 < -\left.\frac{dN_2}{dN_1}\right|_{u_1 = \bar{u}_2} = \frac{\beta_1 w_1}{\beta_2 w_2 + c_1(\beta_1 w_1 N_1 + \beta_2 w_2 N_2)/(C_1 + N_2)}$$

$$< \frac{\beta_1 w_1}{\beta_2 w_2} < -\left.\frac{dN_2}{dN_1}\right|_{u_2 = \bar{u}_2} = \frac{\beta_1 w_1 + c_2(\beta_1 w_1 N_1 + \beta_2 w_2 N_2)/(C_2 + N_1)}{\beta_2 w_2} \tag{23}$$

for any $N_1 > 0$ and $N_2 > 0$. This means that in Fig. 5.1 the curve representing

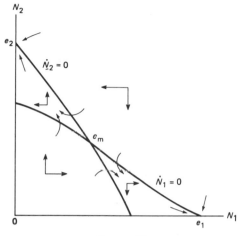

FIGURE 5.1

$\dot{N}_2 = 0$ is always steeper than the curve representing $\dot{N}_1 = 0.$[5] Then, it follows from condition (13) that whenever there exists a mixed-city equilibrium, say e_m in Fig. 5.1, there must be two other equilibrium points with $N_i^* > 0$ and $N_j^* = 0$ $(i, j = 1, 2)$, say e_1 and e_2 in Fig. 5.1. It also follows from the adjustment process (14) that, as shown in Fig. 5.1, the mixed-city equilibrium e_m is always globally unstable and the city will eventually approach either e_1 or e_2, namely an "all black" or "all white" city with

$$N_i^* = (\alpha_i)^{1/b_i}(\beta_i)^{-1}(w_i)^{-1}(C_i)^{-c_i/b_i}(u_i)^{-1/b_i}, \qquad N_j^* = 0 \qquad (i, j = 1, 2). \quad (24)$$

Thus we have established the following theorem.

Theorem 5.2 *In the Cobb–Douglas case with negative intergroup externalities, the mixed-city equilibrium is globally unstable and the city will eventually accommodate only one group of households.*

5.3   Positive Intragroup Externalities

In this section, there are assumed to be no externalities between the two groups as in the previous sections, but some positive externalities within each group. In other words, the two groups neither like or dislike each other, but each group likes its own in the sense that the utility of a typical household

---

[5] Incidentally, this ensures the uniqueness of a mixed-city equilibrium.

in one group tends to increase as the size of its own group increases. This kind of neighborhood externality can be called positive intragroup externality and is expressed as

$$U_i = F_i(d_i, h_i)E_i(N_i) \qquad (i = 1, 2), \tag{25}$$

where

$$E_i'(N_i) > 0 \qquad (i = 1, 2). \tag{26}$$

Following the terminology adopted by Schelling (1971), we may call the utility function (25) "congregationist preferences," meaning that each household wants to congregate with its own group but is indifferent to the presence of the other group of households in the city. Although congregationist preferences seem less discriminatory than the type of preferences we have assumed in the negative intergroup externality case, it turns out that their long-run effect on the residential composition of the city is equally destabilizing. It is interesting to note that such a result has been suggested by Schelling (1971) in connection with his simple experiments on segregation.

In the present case, the indirect utility function becomes

$$V_i = G_i(r, w_i)E_i(N_i) \qquad (i = 1, 2). \tag{27}$$

At each moment of time with N_i given, the utility level for group i is written as

$$u_i(N_1, N_2) = G_i[r(N_1, N_2), w_i]E_i(N_i) \qquad (i = 1, 2), \tag{28}$$

where r is given by Eq. (10). Then we can prove the following instability theorem.

Theorem 5.3 *The mixed-city equilibrium is locally unstable in the presence of positive intragroup externalities, given that residential land is a non-Giffen good for every household.*

Proof   In a small neighborhood of the mixed-city equilibrium, we have system (16) with

$$
\begin{aligned}
a_{11} &\equiv f_1'(0)[(\partial G_1/\partial r)(\partial r/\partial N_1)E_1 + G_1 E_1'(N_1)], \\
a_{12} &\equiv f_1'(0)(\partial G_1/\partial r)(\partial r/\partial N_2)E_1, \\
a_{21} &\equiv f_2'(0)(\partial G_2/\partial r)(\partial r/\partial N_1)E_2, \\
a_{22} &\equiv f_2'(0)[(\partial G_2/\partial r)(\partial r/\partial N_2)E_2 + G_2 E_2'(N_2)],
\end{aligned}
\tag{29}
$$

where all the variables are evaluated at the equilibrium. In this case, the stability condition that $a_{11} + a_{22} < 0$ and $a_{11}a_{22} - a_{12}a_{21} > 0$ cannot be satisfied, since

$$a_{11}a_{22} - a_{12}a_{21} = f'_1(0)f'_2(0)\{(\partial G_1/\partial r)(\partial r/\partial N_1)E_1 G_2 E'_2(N_2)$$
$$+ [(\partial G_2/\partial r)(\partial r/\partial N_2)E_2 + G_2 E'_2(N_2)]G_1 E'_1(N_1)\}$$
$$= f'_1(0)f'_2(0)\{(\partial G_2/\partial r)(\partial r/\partial N_2)E_2 G_1 E'_1(N_1)$$
$$+ [(\partial G_1/\partial r)(\partial r/\partial N_1)E_1 + G_1 E'_1(N_1)]G_2 E'_2(N_2)\} < 0,$$

if we have $a_{11} < 0$ and/or $a_{22} < 0$, which is implied by $a_{11} + a_{22} < 0$,   Q.E.D.

Finally, it should be added that in the presence of *both* negative intergroup externalities *and* positive intragroup externalities, the instability of the mixed-city equilibrium holds a fortiori, as can easily be shown. A special case of this extended model is the case with externalities depending on the ratio of group sizes $E_i(N_i/N_j)$, where $E'_i(\ ) > 0$; a form which Schelling (1969, 1971) has adopted in this simple analysis of segregation.

5.4   A Monocentric City Case

So far we have assumed that land is perfectly homogeneous with no differential transport cost incurred within the city. This assumption does not hold, however, in the standard model of a monocentric city with positive transport cost incurred by every resident in commuting from his residence to work in the central business district (CBD). In this case, as seen in the previous chapters, those pieces of land which are closer to the CBD have locational advantages in terms of transport cost over those pieces of land which are located farther from the CBD. Then the question is how the instability results we have obtained in the previous cases will be modified.

As shown in the previous chapters, in this kind of model with more than one class of households, a segregated pattern of residence will emerge in equilibrium; the whole land area will be subdivided into the same number of zones as the number of household classes, and each zone is exclusively occupied by a group of households in the same class. If, in particular, there are two groups of households having identical transport cost functions and identical preferences for a consumption good and residential land, but having different income levels, then the lower-income group can be shown to occupy the zone closer to the CBD, while consuming less land than the higher-income group (see e.g., Alonso, 1964; Muth, 1969). It turns out that spatial segregation of this type will have a certain effect on the long-run dynamic property of a mixed city with neighborhood externalities. In fact, its effect is stabilizing and, as a result, the system may be stable or unstable, depending on the magnitude of the destabilizing effect of neighborhood externalities relative to the stabilizing effect of spatial segregation within the city.

For the purpose of illustration, it is sufficient to consider the case of negative intergroup externalities with the Cobb–Douglas utility function,

$$U_i = (d_i)^a (h_i)^b (C_i + N_j)^{-c_i}, \qquad i = j \qquad (i, j = 1, 2), \qquad (30)$$

where a, b, c_i, and C_i ($i = 1, 2$) are all positive constants. Furthermore, we assume that the transport cost t incurred by a household living at distance x from the CBD can be written as $t(x) = qx$, where marginal transport cost q is a given positive constant and the same for the two groups. Since all households in the same group must attain the same utility level regardless of their location in equilibrium, we find

$$u_i = A[r_i(x)]^{-b} (w_i - qx)^{a+b} (C_i + N_j)^{-c_i}, \qquad i \neq j \qquad (i, j = 1, 2), \quad (31)$$

where the expression on the right-hand side is the indirect utility function with A and $w_i - qx$ in place of A_i and w_i, respectively, in Eq. (19), where $A = a^a b^b (a + b)^{-(a+b)}$. From Eq. (31), group i's bid rent $r_i(x)$ at x corresponding to a utility level u_i which is common for all households in group i can be expressed as

$$r_i(x) = A^{1/b} (w_i - qx)^{(a+b)/b} (C_i + N_j)^{-c_i/b} (u_i)^{-1/b} \qquad (i = 1, 2). \quad (32)$$

Without loss of generality, let us assume that $w_1 < w_2$. As already pointed out, group 1 will be accommodated in the inner zone, i.e., the land area between the CBD and the residential boundary between groups 1 and 2. In view of the fact that the reciprocal of the amount of land per household in group 1 at x, $1/h_1(x)$, is equal to the density of households in group 1 at x, we have

$$\int_0^{x_1} \frac{s(x)}{h_1(x)} \, dx = N_1, \qquad (33)$$

where $s(x)$ is the amount of residential land available at distance x from the CBD,[6] and x_1 is the distance from the CBD to the boundary between the two groups. Similarly, all households in group 2 will be located in the outer zone, i.e., the area between x_1 and the outer city boundary x_2,

$$\int_{x_1}^{x_2} \frac{s(x)}{h_2(x)} \, dx = N_2. \qquad (34)$$

The residential boundary x_1 is endogenously determined by the condition

---

[6] Note that $s(x) = 1$ in the linear city case and $s(x) = 2\pi x$ in the circular city case. In what follows we assume that $s(x) > 0$ for all $0 \leq x \leq x_2$.

that the overall rent function should be continuous at x_1,

$$r_1(x_1) = r_2(x_1), \tag{35}$$

whereas the outer city boundary x_2 is given by $r_2(x_2) = 0$, i.e., $x_2 = w_2/q$ in view of Eq. (32), where the opportunity cost of land is assumed to be zero for the sake of simplicity.

In order to determine u_1 and u_2 as functions of N_1 and N_2, we first solve Eq. (35) for x_1, using Eq. (32), as

$$x_1 = (w_2 T_2 - w_1 T_1)/[q(T_2 - T_1)] \equiv x_1(u_1, u_2, N_1, N_2), \tag{36}$$

where

$$T_i \equiv (C_i + N_j)^{-c_i/(a+b)}(u_i)^{-1/(a+b)} \qquad (i = 1, 2).$$

For our later use we note that

$$
\begin{aligned}
\frac{\partial x_1}{\partial u_1} &= \frac{-w_1(T_2 - T_1) + w_2 T_2 - w_2 T_1}{q(T_2 - T_1)^2}\frac{\partial T_1}{\partial u_1} = -\frac{Y}{u_1}, \\
\frac{\partial x_1}{\partial u_2} &= \frac{w_2(T_2 - T_1) - (w_2 T_2 - w_1 T_1)}{q(T_2 - T_1)^2}\frac{\partial T_2}{\partial u_2} = \frac{Y}{u_2}, \\
\frac{\partial x_1}{\partial N_1} &= \frac{w_2(T_2 - T_1) - (w_2 T_2 - w_1 T_1)}{q(T_2 - T_1)^2}\frac{\partial T_2}{\partial N_1} = \frac{c_2 Y}{C_2 + N_1}, \\
\frac{\partial x_1}{\partial N_2} &= \frac{-w_1(T_2 - T_1) + w_2 T_2 - w_1 T_1}{q(T_2 - T_1)^2}\frac{\partial T_1}{\partial N_2} = -\frac{c_1 Y}{C_1 + N_2},
\end{aligned}
\tag{37}
$$

where

$$Y \equiv \frac{(w_2 - w_1)T_1 T_2}{(a + b)q(T_2 - T_1)^2} > 0.$$

Since the demand for land per household in group i at x is

$$h_i(x) = \beta(w_i - qx)/r_i(x) = B(w_i - qx)^{-a/b}(C_i + N_j)^{c_i/b}(u_i)^{1/b}$$

with $\beta \equiv b/(a + b)$ and $B \equiv \beta A^{-1/b}$, it follows from Eqs. (33) and (34) that

$$M_1(u_1, N_2, x_1) \equiv B^{-1}(C_1 + N_2)^{-c_1/b}(u_1)^{-1/b}\int_0^{x_1}(w_1 - qx)^{a/b}s(x)\,dx = N_1,$$

$$\tag{38}$$

$$M_2(u_2, N_1, x_1) \equiv B^{-1}(C_2 + N_1)^{-c_2/b}(u_2)^{-1/b}\int_{x_1}^{x_2}(w_2 - qx)^{a/b}s(x)\,dx = N_2.$$

From Eqs. (36) and (38) we can determine u_1 and u_2 as functions of N_1 and N_2, and calculate $\partial u_i/\partial N_j$ $(i, j = 1, 2)$ as follows.

Let $Z_i \equiv \partial M_i/\partial x_1 (i = 1, 2)$. It is easy to see from Eq. (38) that $Z_1 > 0$ and

$Z_2 < 0$. Total differentiation of system (38) with respect to u_1, u_2, N_1, and N_2 gives

$$\begin{bmatrix} H & J \\ K & L \end{bmatrix} \begin{bmatrix} du_1 \\ du_2 \end{bmatrix} = \begin{bmatrix} P & Q \\ R & S \end{bmatrix} \begin{bmatrix} dN_1 \\ dN_2 \end{bmatrix}, \tag{39}$$

where

$$H \equiv -\frac{N_1}{bu_1} + Z_1 \frac{\partial x_1}{\partial u_1} = -\frac{Z_1 Y + N_1/b}{u_1}, \qquad J \equiv Z_1 \frac{\partial x_1}{\partial u_2} = \frac{Z_1 Y}{u_2},$$

$$K \equiv Z_2 \frac{\partial x_1}{\partial u_1} = -\frac{Z_2 Y}{u_1}, \qquad L \equiv -\frac{N_2}{bu_2} + Z_2 \frac{\partial x_1}{\partial u_2} = \frac{Z_2 Y - N_2/b}{u_2},$$

$$P \equiv 1 - Z_1 \frac{\partial x_1}{\partial N_1} = 1 - \frac{c_2 Z_1 Y}{C_2 + N_1},$$

$$Q \equiv \frac{c_1 N_1}{b(C_1 + N_2)} - Z_1 \frac{\partial x_1}{\partial N_2} = \frac{c_1(Z_1 Y + N_1/b)}{C_1 + N_2}, \tag{40}$$

$$R \equiv \frac{c_2 N_2}{b(C_2 + N_1)} - Z_2 \frac{\partial x_1}{\partial N_1} = -\frac{c_2(Z_2 Y - N_2/b)}{C_2 + N_1},$$

$$S \equiv 1 - Z_2 \frac{\partial x_1}{\partial N_2} = 1 + \frac{c_1 Z_2 Y}{C_1 + N_2}.$$

By setting $dN_2 = 0$ and using Cramer's rule, we can solve system (39) for $\partial u_1/\partial N_1$ and $\partial u_2/\partial N_1$ as

$$\frac{\partial u_1}{\partial N_1} = \frac{1}{D} \begin{vmatrix} P & J \\ R & L \end{vmatrix} = \frac{1}{D} \left[\left(1 - \frac{c_2 Z_1 Y}{C_2 + N_1} \right) \frac{Z_2 Y - N_2/b}{u_2} \right.$$

$$\left. + \frac{Z_1 Y}{u_2} \frac{c_2(Z_2 Y - N_2/b)}{C_2 + N_1} \right]$$

$$= \frac{Z_2 Y - N_2/b}{D u_2},$$

$$\frac{\partial u_2}{\partial N_1} = \frac{1}{D} \begin{vmatrix} H & P \\ K & R \end{vmatrix} = \frac{1}{D} \left[\frac{Z_1 Y + N_1/b}{u_1} \frac{c_2(Z_2 Y - N_2/b)}{C_2 + N_1} \right. \tag{41}$$

$$\left. + \left(1 - \frac{c_2 Z_1 Y}{C_2 + N_1} \right) \frac{Z_2 Y}{u_1} \right]$$

$$= \frac{1}{D} \left[\frac{Z_2 Y}{u_1} - \frac{c_2(N_1 N_2 + b Z_1 N_2 Y - b Z_2 N_1 Y)}{u_1 b^2(C_2 + N_1)} \right]$$

$$= \frac{b^2 Z_2 Y - c_2 N_2(1 + g)/(1 + C_2/N_1)}{D u_1 b^2},$$

where

$$D \equiv HL - JK = N_1 N_2 (1 + g)/(u_1 u_2 b^2) > 0 \qquad (42)$$

and

$$g \equiv bY[(Z_1/N_1) - (Z_2/N_2)] > 0. \qquad (43)$$

Similarly, by setting $dN_1 = 0$, we obtain

$$\frac{\partial u_1}{\partial N_2} = \frac{1}{D} \begin{vmatrix} Q & J \\ S & L \end{vmatrix} = \frac{-b^2 Z_1 Y - c_1 N_1 (1 + g)/(1 + C_1/N_2)}{D u_2 b^2},$$

$$\frac{\partial u_2}{\partial N_2} = \frac{1}{D} \begin{vmatrix} H & Q \\ K & S \end{vmatrix} = -\frac{Z_1 Y + N_1/b}{D u_1}. \qquad (44)$$

Now we can examine the stability property of the dynamic system (14) in the monocentric city case. We prove the following theorem.

Theorem 5.4 *In the monocentric city case with the Cobb–Douglas utility function (30) involving negative intergroup externalities, the mixed-city equilibrium is locally stable if*

$$c_i/(1 + C_i/N_j) < b/(1 + g) \qquad (i = 1, 2), \qquad (45)$$

and it is locally unstable if

$$c_i/(1 + C_i/N_j) > b/(1 + g) \qquad (i = 1, 2). \qquad (46)$$

Proof In view of (41) and (44), we have the linear system (16) with

$$a_{11} \equiv f'_1(0)\, \partial u_1/\partial N_1 < 0, \qquad a_{12} \equiv f'_1(0)\, \partial u_1/\partial N_2 < 0,$$

$$a_{21} \equiv f'_2(0)\, \partial u_2/\partial N_1 < 0, \qquad a_{22} \equiv f'_2(0)\, \partial u_2/\partial N_2 < 0. \qquad (47)$$

Furthermore, by rearranging terms, we find

$$\begin{aligned}
a_{11}a_{22} - a_{12}a_{21} = \frac{f'_1(0)f'_2(0)(1 + g)}{D^2 u_1 u_2 b^2} & \left[Z_1 Y N_2 \left(\frac{b}{1 + g} - \frac{c_2}{1 + C_2/N_1} \right) \right. \\
& - Z_2 Y N_1 \left(\frac{b}{1 + g} - \frac{c_1}{1 + C_1/N_2} \right) \\
& \left. + \frac{(1 + g)N_1 N_2}{b^2} \left(\frac{b^2}{(1 + g)^2} - \frac{c_1}{1 + C_1/N_2} \frac{c_2}{1 + C_2/N_1} \right) \right].
\end{aligned}$$

$$(48)$$

Then, it is clear that expression (48) is positive if condition (45) is met, and it is negative if (46) is satisfied.   Q.E.D.

A possible economic interpretation of conditions (45) and (46) is as follows. Suppose initially u_1 is smaller than $\bar{u}_1$. The resultant out-migration of group

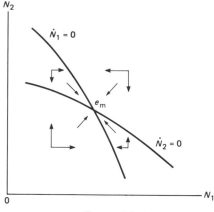

FIGURE 5.2

1 will make the city more attractive to the other group and thus induce more in-migration (or less out-migration) of group 2. This in turn tends to lower u_1 due to negative intergroup externalities, and u_1 will diverge from $\bar{u}_1$ through time. On the other hand, owing to spatial segregation, a decrease in N_1 tends to raise the utility level for group 1 by increasing the amount of land to be occupied by each household in group 1, although this effect is partially offset by the inward movement of the residential boundary between the two groups as a result of a decline in the land rent paid by a smaller number of households in group 1. Thus, the mixed-city equilibrium is stable or unstable as the former destabilizing effect represented by $c_i/(1 + C_i/N_j)$, i.e., the elasticity of U_i with respect to N_j, is smaller or greater than the latter stabilizing effect $b/(1 + g)$, i.e., the elasticity of U_i with respect to h_i. Note that the presence of the "discount factor" $(1 + g)$ is due to the offsetting effect of the boundary movement. Figures 5.2 and 5.3 illustrate stable and

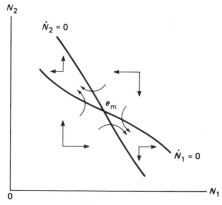

FIGURE 5.3

unstable cases, respectively. It should be noted that a similar result can be obtained in the case of positive intragroup externalities.

5.5 Final Remarks

We have shown that in the presence of negative intergroup externalities or positive intragroup externalities, (1) the mixed-city equilibrium is always unstable if land is perfectly homogeneous in the city with no differential transport cost incurred so that there is no spatial segregation for economic reasons, and (2) the mixed-city equilibrium is not necessarily unstable in the case of a monocentric city with positive transport cost where spatial segregation occurs within the city. In fact, in the latter case with a Cobb–Douglas utility function, the equilibrium is locally stable if the degree of externalities is sufficiently small relative to the elasticity of utility with respect to land.

A few remarks are now in order. First, the present analysis can be easily extended to include more complicated cases with any mixture of four kinds of neighborhood externalities, namely, positive intergroup, negative intergroup, positive intragroup, and negative intragroup. It seems obvious that positive intergroup and negative intragroup externalities have some stabilizing effects, while negative intergroup and positive intragroup externalities are destabilizing as we have analyzed in this chapter. If, for instance, group 1 likes group 2 who dislikes group 1, the mixed-city equilibrium is stable or unstable, depending on whether the degree of positive intergroup externalities on the part of group 1 is higher or lower than the degree of negative intergroup externalities on the part of group 2.

Second, we have dealt only with a kind of segregation which results from the economic behavior of the household choosing more (less) land in exchange for higher (lower) transport cost. However, spatial segregation is expected to be always stabilizing irrespective of its cause or reason. Thus, it could be conjectured that the (in)stability results we have obtained in the monocentric city case hold true if spatial segregation within the city results directly from negative intergroup externalities themselves, although a more complicated model would be needed to derive a segregated residential pattern on the basis of neighborhood externalities.

Finally, it should be pointed out that the phenomenon of increasing returns in production can be treated in essentially the same way as positive intragroup externalities if increasing returns take such a form that the income level (wage rate) for each group is an increasing function of its own group size due to increasing marginal labor productivity. Obviously, in the presence of increasing returns of this kind, the mixed-city equilibrium tends to be unstable: this may appear to be a dynamic version of the well-known proposition of complete specialization in the presence of increasing returns in international trade theory. What is interesting here, though, is that even in the presence of increasing returns, as our analysis suggests, the city may not completely specialize in the long run if there is spatial segregation between two groups of households (workers) in the city and if the degree of increasing returns is sufficiently small relative to the elasticity of utility with respect to residential land.

REFERENCES

Alonso, W. (1964). *Location and Land Use.* Harvard Univ. Press, Cambridge, Massachusetts.
Anas, A. (1980). A model of residential change and neighborhood tipping. *Journal of Urban Economics* **7,** 358–370.
Miyao, T. (1978). Dynamic instability of a mixed city in the presence of neighborhood externalities. *American Economic Review* **68,** 454–463.
Muth, R. F. (1969). *Cities and Housing.* Univ. of Chicago Press, Chicago.
Rose-Ackerman, S. (1975). Racism and urban structure. *Journal of Urban Economics* **2,** 85–103.
Schelling, T. C. (1969). Models of segregation. *American Economic Review* **59,** 488–493.
Schelling, T. C. (1971). Dynamic models of segregation. *Journal of Mathematical Sociology* **1,** 143–186.
White, M. J. (1977). Urban models of race discrimination. *Regional Science and Urban Economics* **7,** 217–232.
Yinger, J. (1976). Racial prejudice and racial residential segregation in an urban model. *Journal of Urban Economics* **3,** 383–396.

CHAPTER **6**

Probabilistic Approach
to Neighborhood Choice

In this chapter we adopt a probabilistic choice approach to the problem of neighborhood choice in a city and develop a general model of probabilistic location choice by many types of individuals (households) who interact among themselves in the presence of neighborhood effects.[1] The model we develop here focuses on intergroup externalities, but ignores individual preferences for space and other commodities in order to keep the model manageable enough to derive some meaningful results. As such, our model may be viewed as a generalization of Schelling's original model of segregation between two groups into a situation with many types of individuals choosing among many alternative locations or neighborhoods in a city [Schelling, 1969, 1971; see especially his bounded-neighborhood model (1971, pp. 167–186)].

By applying the probabilistic theory of qualitative choice, we suppose that the selection probabilities of individuals of each type for various locations are functions of the representative utilities obtained from all possible locations (Marschak, 1960; McFadden, 1974). Neighborhood effects are taken into account by assuming that the representative utility obtained by individuals of each type is a function of the numbers of individuals of all types who are actually choosing the same location. Note that in this chapter

---

[1] This chapter draws on Miyao (1978).

we deal with neighborhood effects within each location (neighborhood), but not across locations.

In the present context, equilibrium is defined as an allocation of individuals among locations such that for individuals of every type the selection probability for each location is equal to the proportion of individuals actually choosing the same location. The existence of equilibrium is guaranteed under some continuity assumptions. The equilibrium is shown to be unique and dynamically stable in the presence of neighborhood effects, provided that the degree of neighborhood effects is sufficiently small.

Finally, we consider a simple but important case with two types of individuals (e.g., blacks and whites) and two locations (e.g., an inner city and its suburban neighborhood), and show that if the two groups do not like each other, the mixed-city equilibrium is stable *or* unstable as the degree of negative externalities between them is small *or* large relative to the variance of the distribution of utility over individuals of each type.

6.1   The Model and Existence

Each individual is assumed to choose among n alternative locations, which can be considered as n different neighborhoods in a city. There are m types of individuals with a certain number of individuals for each type, where any two different types can be easily distinguished by observation.

For type k $(k = 1, \ldots, m)$, the utility obtained from location i, say U^{ki}, is assumed to be randomly distributed over the population of individuals of type k, and the selection probability of type k for location i, say P^{ki}, can be defined as the probability that an individual drawn randomly from the population of type k will find location i most attractive,

$$P^{ki} = \text{Prob}(U^{ki} > U^{kj} \text{ for all } j \neq i) \qquad (k = 1, \ldots, m; \quad i = 1, \ldots, n). \quad (1)$$

Under certain assumptions (see McFadden, 1974, p. 108), P^{ki} can be expressed as a function of "representative utilities," $u^{ki}, \ldots, u^{kn}$ $(k = 1, \ldots, m)$, where u^{kj} is the representative utility of type k for location j and is equal to the non-stochastic component of U^{kj}. Then we have

$$P^{ki}(u^{ki}, \ldots, u^{kn}) \geq 0 \qquad \text{and} \qquad \sum_{i=1}^{n} P^{ki}(u^{ki}, \ldots, u^{kn}) = 1$$

$$\text{for all} \quad u^{ki}, \ldots, u^{kn} \qquad (k = 1, \ldots, m). \quad (2)$$

Let us consider "neighborhood effects" resulting from the interaction of various types of individuals at each location. Denoting by N_i^k the number of individuals of type k who actually choose location i, we assume that the representative utility of type k for location i, u^{ki}, depends on the numbers of individuals of all types at location i,

$$u^{ki} = u^{ki}(N_i^1, \ldots, N_i^m) \qquad (k = 1, \ldots, m; \quad i = 1, \ldots, n). \qquad (3)$$

If individuals of type k like (or dislike) those of type j at location i, then $\partial u^{ki}/\partial N_i^j > 0$ (or < 0). Increasing returns (agglomeration economies) and decreasing returns (congestion) can be represented by the derivative of $u^{ki}(N_i^1 + \cdots + N_i^m)$ being positive and negative, respectively.

In the present context, we define the concept of equilibrium as an allocation of individuals among n alternative locations such that the selection probability of type k for location i, P^{ki}, is equal to the proportion of individuals of type k who actually choose location i, i.e., N_i^k/L^k, where L^k is the total number of individuals of type k, which is given as a positive constant. Here we treat the number of individuals as a continuous variable in order to avoid any complexities arising from a discrete number of individuals. Thus, equilibrium is defined as an allocation of individuals $N_i^k \geq 0$ such that

$$\sum_{i=1}^{n} N_i^n = L^k > 0 \qquad (k = 1, \ldots, m), \qquad (4)$$

$$N_i^k = P^{ki}\big[u^{ki}(N_1^1, \ldots, N_1^m), \ldots, u^{kn}(N_n^1, \ldots, N_n^m)\big]L^k$$

$$(k = 1, \ldots, m; \quad i = 1, \ldots, n). \qquad (5)$$

Then we can prove the existence of equilibrium.

Theorem 6.1   *There exists an equilibrium, if $u^{ki}(N_i^1, \ldots, N_i^m)$ is continuous in N_i^h for $0 \leq N_i^h \leq L \equiv \sum_{k=1}^{m} L^k$ ($h = 1, \ldots, m$), and $P^{ki}(u^{k1}, \ldots, u^{kn})$ is continuous in $u^{k1}, \ldots, u^{kn}$ and satisfies property (2) for all $k = 1, \ldots, m$ and all $i = 1, \ldots, n$.*

Proof   Consider a simplex defined by

$$N_i^k \geq 0 \qquad \text{and} \qquad \sum_k \sum_i N_i^k = L > 0. \qquad (6)$$

Then we have a continuous function $P^{ki}L^k$, which maps from the simplex (6) to itself. Thus, Brouwer's fixed-point theorem ensures the existence of a fixed point $N_i^{k*} \geq 0$ such that[2]

$$N_i^{k*} = P^{ki}\big[u^{k1}(N_1^{1*}, \ldots, N_1^{m*}), \ldots, u^{kn}(N_n^{1*}, \ldots, N_n^{m*})\big]L^k$$

for all k and all i. Equation (4) is also satisfied, since $\sum_i N_i^{k*} = \sum_i P^{ki}(\ )L^k = L^k$ in view of (2).   Q.E.D.

---

[2] For an excellent presentation and proof of Brouwer's fixed-point theorem, see Nikaido (1970, pp. 303–308).

6.2   Uniqueness

Let us further assume that for each type k, P^{ki} will decrease (or not increase) when any one of u^{kj}'s other than u^{ki} increases,

$$P^{ki}_j \equiv \partial P^{ki}/\partial u^{kj} \leqq 0 \qquad (i = j). \tag{7}$$

This means that all locations are substitutes for (or independent of) each other, and location i becomes less (or no more) attractive compared to location j whose utility increases. It is easy to see that property (7) implies that

$$P^{ki}_i \equiv \partial P^{ki}/\partial u^{ki} = -\sum_{j \neq i}(\partial P^{kj}/\partial u^{ki}) \geqq 0 \qquad (k = 1, \ldots, m; \quad i = 1, \ldots, n), \tag{8}$$

because $\sum_{j=1}^{n}(\partial P^{kj}/\partial u^{ki}) = 0$ from property (2).[3]
Define $(m \times n)$-dimensional vectors N and f as

$$N \equiv (N^1_1, \ldots, N^1_n, N^2_1, \ldots, N^2_n, \ldots, N^m_1, \ldots, N^m_n), \tag{9}$$

$$f \equiv (f^{11}, \ldots, f^{1n}, f^{21}, \ldots, f^{2n}, \ldots, f^{m1}, \ldots, f^{mn}), \tag{10}$$

where

$$f^{ki} \equiv N^k_i - P^{ki}(\ )L^k \qquad (k = 1, \ldots, m; \quad i = 1, \ldots, n. \tag{11}$$

Then it is clear that the equilibrium N^* is given by solving the system $f(N) = 0$. Also define

$$u^{ki}_h \equiv \partial u^{ki}/\partial N^h_i \qquad (k, h = 1, \ldots, m; \quad i = 1, \ldots, n). \tag{12}$$

We shall prove that the equilibrium is unique if the "degree" of externality is sufficiently small.

Theorem 6.2   *The solution to the equilibrium system $f(N) = 0$ is unique, if $P^{ki}(u^{k1}, \ldots, u^{kn})$ satisfies properties (2) and (7) for all $u^{k1}, \ldots, u^{kn}$ $(i = 1, \ldots, m;$ $i = 1, \ldots, n)$ and the degree of externality $|u^{kj}_h|$ is small enough to satisfy*

$$\sum_{k=1}^{m} (P^{kj}_j |u^{kj}_h| L^k) < \tfrac{1}{2} \qquad for \quad 0 \leqq N^k_i \leqq L^k$$

$$(h = 1, \ldots, m; \quad i, j = 1, \ldots, n). \tag{13}$$

Proof   Since P^{ki} and u^{ki} are all differentiable, we have a differentiable mapping $f: \Omega \to R^{m \times n}$, where Ω is a closed rectangular region $\Omega = \{N|0 \leqq N^k_i \leqq L^k\}$. The Jacobian matrix of the mapping f is

$$J(N) = I + A \tag{14}$$

---

[3] Properties.(7) and (8) are actually satisfied by McFadden's (1974) selection probabilities $P^{ki} = e^{u^{ki}}/(\sum_{j=1}^{n} e^{u^{kj}})$ and also by Marschak's (1960) $P^{ki} = u^{ki}/(\sum_{j=1}^{n} u^{kj})$.

with

$$A = \begin{vmatrix} A_{11} & \cdots & A_{1m} \\ \vdots & A_{kh} & \vdots \\ A_{m1} & \cdots & A_{mn} \end{vmatrix}, \tag{15}$$

where

$$A_{kh} = \begin{bmatrix} -P_1^{k1} u_h^{k1} L^k & \cdots & -P_n^{k1} u_h^{kn} L^k \\ \vdots & -P_j^{ki} u_h^{kj} L^k & \vdots \\ -P_1^{kn} u_h^{k1} L^k & \cdots & -P_n^{kn} u_h^{kn} L^k \end{bmatrix} \qquad (k, h = 1, \ldots, m). \tag{16}$$

It can be shown that $I + A$ is a dominant diagonal matrix with positive diagonal elements for all N in Ω, because for any column, i.e., for any h and j,

$$1 - P_j^{hj} u_h^{hj} L^h \geqq 1 - P_j^{hj} |u_h^{hj}| L^h$$
$$> \sum_{i \neq j} |P_j^{hi}| \, |u_h^{hj}| L^h + \sum_{k \neq h} (\sum_{i=1}^{n} |P_j^{ki}| \, |u_h^{kj}| L^k). \tag{17}$$

where the last inequality follows from (13) as

$$1 > 2P_j^{hj} |u_h^{hj}| L^h + \sum_{k \neq h} (2P_j^{kj} |u_h^{kj}| L^k)$$

$$= P_j^{hj} |u_h^{hj}| L^h + \sum_{i \neq j} |P_j^{hi}| \, |u_h^{hi}| L^h + \sum_{k \neq h} (\sum_{i=1}^{n} |P_j^{hi}| \, |u_h^{kj}| L^k), \tag{18}$$

in view of $P_j^{kj} = \sum_{i \neq j} |P_j^{ki}|$ from (2). Thus, according to Gale and Nikaido (1965), the mapping f is univalent in Ω and, as a result, the solution to $f(N) = 0$ is unique. Q.E.D.

6.3 Stability

Suppose that the system is out of equilibrium such that the number of individuals of type k who like location i best is not equal to the number of individuals of type k who actually choose location i for at least one k and one i. In such a disequilibrium situation, some kind of dynamic adjustment must take place over time. We take up two alternative adjustment processes, i.e., discrete adjustment and continuous adjustment.

First, let us assume a one-period time lag for individuals to adjust their actual behavior to their desired one. Then we may describe the adjustment process by the difference equation system

$$N_i^k(t + 1) = P^{ki}\{u^{ki}[N_1^1(t), \ldots, N_1^m(t)], \ldots, u^{kn}[N_n^1(t), \ldots, N_n^m(t)]\}L^k$$

$$(k = 1, \ldots, m; \quad i = 1, \ldots, n), \qquad (19)$$

where

$$\sum_{i=1}^{n} N_i^k(t) = L^k \qquad (k = 1, \ldots, m; \quad i = 1, \ldots, n). \qquad (20)$$

By expressing $N_n^k(t)$ as

$$N_n^k(t) = L^k - \sum_{i=1}^{n-1} N_i^k(t)$$

from (20), we deal with $m \times (n - 1)$ variables,

$$N \equiv (N_1^1, \ldots, N_{n-1}^1, N_1^2, \ldots, N_{n-1}^2, \ldots, N_1^m, \ldots, N_{n-1}^m). \qquad (21)$$

We can show that the equilibrium N^* is locally stable, if the degree of externality is sufficiently small.

Theorem 6.3   *According to the adjustment process* (19), *the equilibrium is locally stable, if* $P^{ki}(u^{k1}, \ldots, u^{kn})$ *satisfies properties* (2) *and* (7), *and*

$$\sum_{k=1}^{m} (2P_j^{kj}|u_h^{kj}| + P_n^{kn}|u_h^{kn}|)L^k < 1 \qquad (h = 1, \ldots, m; \quad j = 1, \ldots, n-1) \quad (22)$$

at the equilibrium.

Proof   Linearization of (19) around the equilibrium, together with (20), leads to

$$N(t + 1) - N^* = B[N(t) - N^*], \qquad (23)$$

with

$$B = \begin{vmatrix} B_{11} & \cdots & B_{1m} \\ \vdots & B_{kh} & \vdots \\ B_{m1} & \cdots & B_{mm} \end{vmatrix}, \qquad (24)$$

where

$$B_{kh} =$$

$$\begin{bmatrix} (P_1^{k1}u_h^{k1} - P_n^{k1}u_h^{kn})L^k & \cdots & (P_{n-1}^{k1}u_h^{kn-1} - P_n^{k1}u_h^{kn})L^k \\ \vdots & (P_j^{ki}u_h^{kj} - P_n^{ki}u_h^{kn})L^k & \vdots \\ (P_1^{kn-1}u_h^{k1} - P_n^{kn-1}u_h^{kn})L^k & \cdots & (P_{n-1}^{kn-1}u_h^{kn-1} - P_n^{kn-1}u_h^{kn})L^k \end{bmatrix}$$

$$(k, h = 1, \ldots, m). \qquad (25)$$

If λ is a characteristic root of B, we know that (see McKenzie, 1960, p. 49)

$$|\lambda| \leq \max_{h,j}(\sum_{k=1}^{m}\sum_{i=1}^{n-1}|P_j^{ki}u_h^{ki} - P_n^{ki}u_h^{kn}|L^k)$$

$$\leq \max_{h,j}\{\sum_{k=1}^{m}[\sum_{k=1}^{n-1}(|P_j^{ki}|\,|u_h^{kj}| + |P_n^{ki}|\,|u_h^{kn}|)]L^k\}$$

$$= \max_{h,j}[\sum_{k=1}^{m}(2P_j^{kj}|u_h^{kj}| + P_j^{kn}|u_h^{kj}| + P_n^{kn}|u_h^{kn}|)L^k]$$

$$\leq \max_{h,j}[\sum_{k=1}^{m}(2P_j^{kj}|u_h^{kj}| + P_n^{kn}|u_h^{kn}|)L^k] \tag{26}$$

for $h = 1,\ldots,m$ and $j = 1,\ldots,n-1$. Thus it follows from (22) that $|\lambda| < 1$. Q.E.D.

Alternatively, we may treat time as continuous and introduce gradual adjustment through time. A natural counterpart of system (19) in the context of continuous time can be given by the differential equation system

$$\dot{N}_i^k(t) = P^{ki}\{u^{ki}[N_1^1(t),\ldots,N_1^m(t)],\ldots,u^{kn}[N_n^1(t),\ldots,N_n^m(t)]\}L^k$$
$$ - N_i^k(t) \qquad (k = 1,\ldots,m; \quad k = 1,\ldots,n), \tag{27}$$

where the dot denotes differentiation with respect to time. Note that if the initial condition $N_i^k(0)$ satisfies Eq. (20) at time 0, (20) will be satisfied for all $t \geq 0$, since

$$\sum_{i=1}^{n} \dot{N}_i^k(t) = \sum_{i-1}^{n}[P^{ki}(t)L^k - N_i^k(t)] = L^k - \sum_{i-1}^{n} N_i^k(t) = 0$$

$$\text{for all} \quad t \geq 0. \tag{28}$$

It can be shown that system (27) is "quasi-stable" if the degree of externality is sufficiently small. (For the concept of quasi-stability, see Uzawa, 1961, p. 618.)

Theorem 6.4 *System* (27) *is quasi-stable if* $P^{ki}(u^{k1},\ldots,u^{kn})$ *satisfies properties* (2) *and* (7) *for all* $u^{ki},\ldots,u^{kn}$, *and*

$$\sum_{k=1}^{m}(P_j^{kj}|u_h^{kj}| + P_n^{kn}|u_h^{kn}|)L^k < 1 \qquad (h = 1,\ldots,m; \quad j = 1,\ldots,n-1) \tag{29}$$

for all $N_i^k \geq 0$ *such that* $\sum_{i=1}^{h} N_i^k = L^k$.

Proof Define a distance function $F(t)$ as

$$
F(t) \equiv \sum_{(k,i)\in I_t^+} \big[P^{ki} \{ u^{k1}[N_1^1(t), \ldots, N_1^m(t)], \ldots,
$$

$$
u^{kn}[N_n^1(t), \ldots, N_n^m(t)] \} L^k - N_i^k(t) \big], \tag{30}
$$

where

$$
I_t^+ \equiv \{ (k,i) \, | \, P^{ki}(t) L^k - N_i^k(t) > 0, \text{ or } P^{ki}(t)L^k - N_i^k(t) = 0
$$

$$
\text{and } d[P^{ki}(t)L^k - N_i^k(t)]/dt > 0 \}.
$$

From the fact that $\sum_{k=1}^m \sum_{i=1}^n (P^{ki}L^k - N_i^k) = 0$, it follows that

$$
F(t) = - \sum_{(k,i)\in I_t^-} \big[P^{ki} \{ u^{k1}[N_1^1(t), \ldots, N_1^m(t)], \ldots,
$$

$$
u^{kn}[N_n^1(t), \ldots, N_n^m(t)] \} L^k - N_i^k(t) \big], \tag{31}
$$

where

$$
I_t^- \equiv \{ (k,i) \, | \, P^{ki}(t) L^k - N_i^k(t) < 0, \text{ or } P^{ki}(t)L^k - N_i^k(t) = 0
$$

$$
\text{and } d[P^{ki}(t)L^k - N_i^k(t)]/dt < 0 \}.
$$

It is clear that $F(t) = 0$ if and only if the system is in equilibrium, and otherwise $F(t) > 0$. In order to show that $F(t)$ will decrease over time, define

$$
G(t) \equiv \overline{\lim_{h \to 0}} \, \frac{F(t+h) - F(t)}{h}.
$$

Then it follows that[4]

$$
G(t) = \sum_{(k,i)\in I_t^+} \frac{d[P^{ki}(t)L^k - N_i^k(t)]}{dt} = \sum_{h=1}^m \sum_{j=1}^n \sum_{(k,i)\in I_t^+} \frac{\partial(P^{ki}L^k - N_i^k)}{\partial N_j^h} \dot{N}_j^h
$$

$$
= \sum_{(h,j)\in I_t^+} \sum_{(k,i)\in I_t^+} \frac{\partial(P^{ki}L^k - N_i^k)}{\partial N_j^h} \dot{N}_j^h - \sum_{(h,j)\in I_t^-} \sum_{(k,i)\in I_t^-} \frac{\partial(P^{ki}L^k - N_i^k)}{\partial N_j^h} \dot{N}_j^h
$$

$$
= \sum_{(h,j)\in I_t^+} \big[\sum_{(k,i)\in I_t^-} (P_j^{ki} u_h^{kj} - P_n^{ki} u_h^{kn})L^k - 1 \big] \dot{N}_j^h
$$

$$
- \sum_{(h,j)\in I_t^-} \big[\sum_{(k,i)\in I_t^-} (P_j^{ki} u_h^{kj} - P_n^{ki} u_h^{kn})L^k - 1 \big] \dot{N}_j^h, \tag{32}
$$

which is negative except in equilibrium, because we have $\dot{N}_j^h > 0$ for at least one $(h,j) \in I_t^+$ and $\dot{N}_j^h < 0$ for at least one $(h,j) \in I_t^-$ out of equilibrium, and the expressions within the parentheses are always negative from (29), since

---

[4] Special care should be taken to find $G(t)$. See Negishi (1972, p. 205).

for $(h,j; k,i) \in I_t^+$ or I_t^-,

$$\sum_{(k,i)} (P_j^{ki} u_h^{kj} - P_n^{ki} u_h^{kn}) L^k \leq \sum_{k=1}^{m} (P_j^{kj} |u_h^{kj}| + P_n^{kn} |u_h^{kn}|) L^k < 1. \qquad (33)$$

Clearly $G(t) = 0$ in equilibrium, because I_t^+ and I_t^- become null sets. Thus, according to Uzawa (1961), system (27) is quasi-stable.   Q.E.D.

It is well known that quasi-stability implies global stability if equilibrium is unique.[5] Thus, Theorems 6.2 and 6.4 yield the following proposition.

Theorem 6.5   *The equilibrium is globally stable, according to process* (27), *if* $P^{ki}(u^{k1}, \ldots, u^{kn})$ *satisfies properties* (2) *and* (7) *for all* $u^{k1}, \ldots, u^{kn}$, *and*

$$\sum_{k=1}^{m} (P_j^{kj} |u_h^{kj}| + P_n^{kn} |u_h^{kn}|) L^k < 1 \qquad for \quad 0 \leq N_i^k \leq L^k$$

$$(h = 1, \ldots, m; \quad i, j = 1, \ldots, n). \qquad (34)$$

6.4   The Two-Type–Two-Location Case

Consider a special case with two types of individuals (say, blacks and whites) choosing between two locations (say, an inner city and its suburb). For individuals of each type, the representative utility is assumed to be positively related to the ratio of the number of individuals of their own type to that of the other type at each location,[6]

$$u^{11} = u^{11}(N_1^1/N_1^2), \qquad u^{12} = u^{12}(N_2^1/N_2^2),$$
$$u^{21} = u^{21}(N_1^2/N_1^1), \qquad u^{22} = u^{22}(N_2^2/N_2^1), \qquad (35)$$

where $u^{ki'}(\ ) > 0$ for all $k, i = 1, 2$.

We focus on the stability analysis and consider the continuous adjustment process (27). In view of the constraints $N_1^1 + N_2^1 = L^1$, $N_1^2 + N_2^2 = L^2$, we find

$$\dot{N}_1^1 = P^{11}\{u^{11}(N_1^1/N_1^2), u^{12}[(L^1 - N_1^1)/(L^2 - N_1^2)]\}L^1 - N_1^1,$$
$$\dot{N}_1^2 = P^{21}\{u^{21}(N_1^2/N_1^1), u^{22}[(L^2 - N_1^2)/(L^1 - N_1^1)]\}L^2 - N_1^2. \qquad (36)$$

---

[5] More generally, if the set of equilibria is countable, quasi-stability is equivalent to global stability. See Uzawa (1961, p. 619).

[6] This is essentially the same assumption as Schelling's (1971) in his bounded-neighborhood model, which focuses on a single neighborhood.

The Jacobian matrix of the system will then be

$$B = \begin{bmatrix} b_{11} & b_{12} \\ b_{21} & b_{22} \end{bmatrix}, \tag{37}$$

where

$$
\begin{aligned}
b_{11} &\equiv \{P_1^{11}u^{11\prime}(1/N_1^2) + P_2^{11}u^{12\prime}[-1/(L^2 - N_1^2)]\}L^1 - 1, \\
b_{12} &\equiv \{P_1^{11}u^{11\prime}[-N_1^1/(N_1^2)^2] + P_2^{11}u^{12\prime}[(L^1 - N_1^1)/(L^2 - N_1^2)^2]\}L^1, \\
b_{21} &\equiv \{P_1^{21}u^{21\prime}[-N_1^2/(N_1^1)^2] + P_2^{21}u^{22\prime}[(L^2 - N_1^2)/(L^1 - N_1^1)^2]\}L^2, \\
b_{22} &\equiv \{P_1^{21}u^{21\prime}(1/N_1^1) + P_2^{21}u^{22\prime}[-1/(L^1 - N_1^1)]\}L^2 - 1.
\end{aligned} \tag{38}
$$

In order to derive a necessary and sufficient condition for stability, let us assume that everything is identical and symmetric at the equilibrium, so that we may write

$$u^{ki\prime} = u', \qquad L^k = L, \qquad P_i^{ki} = P \qquad \text{for} \quad k, i = 1, 2. \tag{39}$$

Then we find

$$b_{11} = 4pu' - 1, \qquad b_{12} = -4pu', \qquad b_{21} = -4pu', \qquad b_{22} = 4pu' - 1, \tag{40}$$

and a necessary and sufficient condition for local stability of the "mixed-city" equilibrium with $N_1^1 = N_2^1 = N_1^2 = N_2^2$ is that $u' < 1/(8p)$.

Theorem 6.6   *In the symmetric two-type–two-location case, the mixed-city equilibrium is locally stable if and only if*

$$u' < 1/(8p), \tag{41}$$

where u' and p are defined as in Eq. (39).

Furthermore, it can be shown that if utilities from the two locations are independently, identically, and logistically distributed, then the expression p is inversely related to the standard deviation of the distribution of utility. Assuming that U^{ki} and U^{k2} are independently and identically distributed, we may write their joint distribution function as the product of marginal probability functions,

$$f(U^{ki})f(U^{k2}) \qquad (k = 1, 2). \tag{42}$$

Then we can find

$$P^{ki} = \text{Prob}(U^{ki} > U^{kj}; j \neq i) = \int_{-\infty}^{\infty} f(U^{ki}) \left[\int_{-\infty}^{U^k_i} f(U^{kj}) \, dU^{kj} \right] dU^{ki}$$

$$(k, i, j = 1, 2). \tag{43}$$

For the sake of illustration, consider the logistic distribution case

$$f(U) = e^{-(U-u)/\tau}/\tau[1 + e^{-(U-u)/\tau}]^2, \tag{44}$$

where

$$\tau \equiv \sqrt{3}\sigma/\pi \quad \text{and} \quad u \equiv E(U). \tag{45}$$

In this case, since everything is symmetric at the equilibrium, it can be proved that[7]

$$p = P_i^{ki} = -\frac{1}{\tau^2} \int_{-\infty}^{\infty} \frac{e^{-(U-u)/\tau} - e^{-2(U-u)/\tau}}{[1 + e^{-(U-u)/\tau}]^4} \, dU$$

$$= \frac{1}{\tau} \frac{\frac{1}{3} - e^{-(U-u)/\tau}}{2[1 + e^{-(U-u)/\tau}]^3} \Big|_{-\infty}^{\infty} = \frac{1}{6\tau} = \frac{\pi}{6\sqrt{3}\sigma}. \tag{46}$$

Combining (41) and (46), we obtain the following theorem.

Theorem 6.7 *In the symmetric two-type–two-location case with utilities distributed independently, identically, and logistically, the mixed-city equilibrium is locally stable if and only if*

$$u' < (3\sqrt{3}/4\pi)\sigma. \tag{47}$$

In other words, the mixed-city equilibrium is locally stable if and only if the degree of externality is small relative to the degree of dispersion of the distribution of utility. Figures 6.1 and 6.2 illustrate stable and unstable cases, respectively.

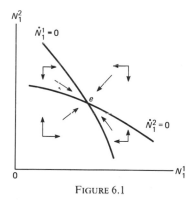

FIGURE 6.1

[7] This manipulation is due to Perry Shapiro.

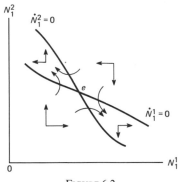

FIGURE 6.2

REFERENCES

Cale, D., and Nikaido, H. (1965). The Jacobian matrix and global univalence of mappings. *Mathematical Annalen* **159,** 81–93.

Marschak, J. (1960). Binary-choice constraints and random utility indicators. In *Mathematical Methods in the Social Sciences* (K. J. Arrow, S. Karlin, and P. Suppes, eds.), pp. 312–329. Stanford Univ. Press, Stanford, California.

McFadden, D. (1974). Conditional logit analysis of qualitative choice behavior. In *Frontiers of Econometrics* (P. Zarembka, ed.), pp. 105–142. Academic Press, New York.

McKenzie, L. (1960). Matrices with dominant diagonals and economic theory. In *Mathematical Methods in the Social Sciences* (K. J. Arrow, S. Karlin, and P. Suppes, eds.), pp. 47–62. Stanford Univ. Press, Stanford, California.

Miyao, T. (1978). A probabilistic model of location choice with neighborhood effects. *Journal of Economic Theory* **19,** 347–358.

Negishi, T. (1972). *General Equilibrium Theory and International Trade.* North-Holland Publ., Amsterdam.

Nikaido, H. (1970). *Introduction to Sets and Mappings in Modern Economics.* North-Holland, Publ., Amsterdam.

Schelling, T. C. (1969). Models of segregation. *American Economic Review* **56,** 488–493.

Schelling, T. C. (1971). Dynamic models of segregation. *Journal of Mathematical Sociology* **1,** 143–186.

Uzawa, H. (1961). The stability of dynamic processes. *Econometrica* **29,** 617–631.

CHAPTER 7

Production and Externality

In this chapter we shall pay attention to another kind of externality, namely, pollution externalities which result from production activities and affect the utility of residents adversely.[1] A typical example of such externalities is air pollution created by a manufacturing industry in a metropolitan area. We examine the existence, uniqueness, and stability of spatial equilibrium in the presence of pollution externalities by setting up an urban location model with both production and residential activities.

In the recent literature, urban location models with pollution externalities have been studied in relation to taxation and zoning policies.[2] Those studies have implicitly assumed the existence, uniqueness, and stability of a spatial equilibrium and have failed to show under what conditions such desirable properties of the equilibrium can be obtained. As suggested by Starrett (1972), there is some reason to believe that in the presence of external diseconomies such as pollution, we may not ensure the existence, uniqueness, or stability of an equilibrium under the usual assumptions.

In order to examine various properties of a spatial equilibrium, we consider a model which is similar to the one developed in Chapter 3, and assume that production functions and utility functions are Cobb–Douglas with the utility of households depending on a consumption good, residential land, total manufacturing output, and the distance from the CBD boundary,

---

[1] This chapter is based on Miyao *et al.* (1980).

[2] See, e.g., Henderson (1977a,b), Hochman (1978), and Stull (1974). For a related study of externalities in relation to hedonic prices, see Polinsky and Shavell (1976).

where the latter two arguments represent the amount and extent of pollution affecting households. It is shown that a spatial equilibrium always exists under the usual conditions. Furthermore, we prove that the equilibrium is unique and dynamically stable, if the degree of externality is sufficiently small relative to the elasticity of utility with respect to residential land. However, the uniqueness and stability of the equilibrium may not be ensured, if the degree of externality is relatively large. In fact, none of the equilibria will be stable for a certain set of speeds of adjustment, if the degree of externality is sufficiently large relative to the elasticity of utility with respect to land.

7.1   The Model with Pollution Externalities

Consider an open city with a single group of identical firms and a single class of identical households. As assumed in Chapter 3, production activities take place in the business area between the central market and the CBD boundary at distance x_b from the market. On the other hand, households are located in the residential area between the CBD boundary and the urban boundary at distance x_c from the market. All products must be transported to the central market where they are sold at an exogenously given price, and households must also travel to the market in order to sell their labor services and to buy consumption goods. Transport (shipping) cost per unit of product q_b and transport (commuting) cost per household q_c are assumed to be functions solely of distance x. The city considered here is open in that both the utility level and the product price are given exogenously from outside the city, whereas the wage rate is determined endogenously within the city.

All firms have identical Cobb–Douglas production functions,

$$Y(x) = T(x)^\alpha N(x)^{1-\alpha} \qquad (0 < \alpha < 1),$$

for a typical firm at distance x from the market, where $Y(x)$, $T(x)$, and $N(x)$ are output, land, and labor, respectively, at distance x. Under perfect competition, we have the zero profit condition, i.e., product price net of transport cost be equal to unit production cost,

$$p - q_b(x) = Ar_b(x)^\alpha w^{1-\alpha}, \tag{1}$$

where p and w are the product price and the wage rate, respectively, $q_b(x)$ and $r_b(x)$ are transport cost and manufacturing bid rent, respectively, at distance x, and A is a positive constant. From Eq. (1) the manufacturing bid rent function can be obtained as

$$r_b(x) = B[p - q_b(x)]^{1/\alpha} w^{-(1-\alpha)/\alpha}, \tag{2}$$

where B is some positive constant.

All households have identical utility functions, depending on residential land h_c, a consumption good d, and the amount of pollution Z, which they receive at their location x,

$$U(x) = U[h_c(x), d(x), Z(x)], \qquad (3)$$

where the partial derivatives of U with respect to h_c, d, and Z are

$$U_h > 0, \qquad U_d > 0, \qquad U_Z < 0.$$

Note that distance does not directly enter the utility function, as we assume away the disutility (or utility) of travel for the sake of simplicity. It may be reasonable to suppose that the amount of pollution received by a household is increasing with total output Y, but is decreasing with the increased distance from the CBD boundary x_b to the location of the household x,

$$Z(x) = Z(Y, t), \qquad Z_Y > 0, \quad Z_t < 0,$$

where $t \equiv x - x_b$. Combining this with Eq. (3), we can write

$$U(x) = U[h_c(x), d(x), Y, t],$$

with $U_Y < 0$, $U_t > 0$.

Let us take up the Cobb–Douglas case with

$$U(x) = h_c(x)^a d(x)^{1-a}(Y_0 + Y)^{-b}(t_0 + t)^c, \qquad 0 < a < 1, \quad b > 0, \quad c > 0, \quad (4)$$

where Y_0 and t_0 are positive constants. These constants are necessary to keep the utility at a finite positive level when $Y = 0$ or $t = 0$, i.e., when there is no pollution or the household is located at the CBD boundary. With the production function being also Cobb–Douglas, we find total output as

$$Y = s \int_0^{x_b} \frac{Y(x)}{T(x)} dx = s \int_0^{x_b} \left[\frac{N(x)}{T(x)} \right]^{1-a} dx,$$

where s is the (constant) amount of land available for production at each distance x. Since profit maximization leads to

$$wN(x)/r_b(x)T(x) = (1 - \alpha)/\alpha$$

for all x, we obtain

$$Y_t = s \int_0^{x_b} \left[\frac{1-\alpha}{\alpha} \frac{r_b(x)}{w} \right]^{1-\alpha} dx = Cw^{-(1-\alpha)/\alpha} \int_0^{x_b} [p - q_b(x)]^{(1-\alpha)/\alpha} dx \qquad (5)$$

in view of Eq. (2), where C is a positive constant. Considering the budget constraint

$$r_c(x)h_c(x) + pd(x) = w - q_c(x),$$

we can write the indirect utility function as

$$u = Dr_c(x)^{-a}[w - q_c(x)](Y_0 + Y)^{-b}(t_0 + t)^c, \tag{6}$$

where D is a positive constant and u is an exogenously given utility level which is common for all households in the city. From Eq. (6), we can determine the residential bid rent function as

$$r_c(x) = Eu^{-1/a}[w - q_c(x)]^{1/a}(Y_0 + Y)^{-b/a}(t_0 + t)^{c/a}, \tag{7}$$

where E is a positive constant.

In the labor market, the excess demand for labor should be zero in equilibrium,

$$s \int_0^{x_b} \frac{N(x)}{T(x)} dx - s \int_{x_b}^{x_c} \frac{1}{h_c(x)} dx = 0. \tag{8}$$

For the land market to be in equilibrium, the overall rent function must be continuous at both the CBD boundary and the urban boundary,

$$r_b(x_b) - r_c(x_b) = 0, \tag{9}$$

$$r_c(x_c) = 0. \tag{10}$$

Note that for the sake of simplicity the opportunity cost of land is assumed to be zero at the urban boundary, as seen in Eq. (10). Furthermore, in equilibrium, the manufacturing bid rent should not be lower (higher, resp.) than the residential bid rent at any point in the business (residential, resp.) area,

$$\begin{aligned} r_b(x) \geq r_c(x) &\quad \text{for all} \quad x \leq x_b, \\ r_b(x) \leq r_c(x) &\quad \text{for all} \quad x_b \leq x \leq x_c. \end{aligned} \tag{11}$$

This condition is necessary for manufacturing activities to take place in the business area and residential activities in the residential area, as assumed at the outset.

7.2 Existence

In order to prove the existence of an equilibrium, we assume that

$$q_b(0) \geq 0, \quad q_c(0) \geq 0, \quad q_b'(x) \geq \epsilon > 0, \quad q_c'(x) \geq \epsilon > 0, \quad \text{for all} \quad x, \tag{12}$$

where ϵ is a positive constant. First, we find from Eqs. (7) and (10) that $w - q_c(x_c) = 0$, or

$$x_c = q_c^{-1}(w) \tag{13}$$

with

$$dx_c/dw = 1/q'_c(x_c) > 0, \qquad (14)$$

and

$$x_c \to 0 \quad \text{as} \quad w \to q_c(0) \geqq 0, \qquad x_c \to \infty \quad \text{as} \quad w \to q_c(\infty) = \infty. \qquad (15)$$

Next, denoting the excess demand for labor in Eq. (8) by F and using Eqs. (2), (5), (7), and (13), we obtain

$$F(w, x_b) = s \int_0^{x_b} \frac{1 - \alpha}{\alpha} \frac{r_b(x)}{w} \, dx - s \int_{x_b}^{x_c} \frac{1}{a} \frac{r_c(x)}{w - q_c(x)} \, dx$$

$$= Jw^{-1/\alpha} \int_0^{x_b} [p - q_b(x)]^{1/\alpha} \, dx$$

$$- K\left[Y_0 + Cw^{-(1-\alpha)/\alpha} \int_0^{x_b} [p - q_b(x)]^{(1-\alpha)/\alpha} \, dx \right]^{-b/a}$$

$$\times \int_{x_b}^{q_c^{-1}(w)} [w - q_c(x)]^{(1-a)/a}(t_0 + x - x_h)^{c/a} \, dx, \qquad (16)$$

where J and K are some positive constants. By partially differentiating $F(w, x_b)$ with respect to w and x_b, it is easy to see from Eq. (16) that

$$\partial F/\partial w < 0 \quad \text{and} \quad \partial F/\partial x_b > 0, \qquad (17)$$

and therefore

$$(dw/dx_b)\big|_{F=0} = -(\partial F/\partial x_b)/(\partial F/\partial w) > 0, \qquad (18)$$

which means that in the (x_b, w) plane the curve representing $F(w, x_b) = 0$ is continuous and upward sloping, as illustrated in Fig. 7.1. We can also show the following. As $x_b \to 0$, we must have $x_b [= q_c^{-1}(w)] \to 0$ to satisfy $F = 0$, because it would follow from (17) that $F < 0$ if $x_c > 0 = x_b$. Therefore, $q_c^{-1}(w) \to 0$, or in other words,

$$w \to q_c(0) \geqq 0 \quad \text{as} \quad x_b \to 0. \qquad (19)$$

Assuming that

$$p > q_b(0), \qquad (20)$$

we can find a positive number, say $\bar{x} > 0$, such that

$$p - q_b(\bar{x}) = 0, \qquad (21)$$

in view of the assumption (12). Then, in order to satisfy $F = 0$, we should have $x_c > x_b$ as $x_b \to \bar{x}$, because Eq. (16) would lead to $F > 0$ if $x_c = x_b = \bar{x} > 0$; i.e., $x_c = q_c^{-1}(w) > x_b = \bar{x}$, or

$$w > q_c(\bar{x}) \quad \text{as} \quad x_b \to \bar{x}. \qquad (22)$$

Considering properties (19) and (22) in addition to (18), we can illustrate the curve representing $F(w, x_b) = 0$ as in Fig. 7.1.

Turning to Eq. (9) and denoting the rent differential $r_b(x_b) - r_c(x_b)$ by G, we obtain from Eqs. (2), (5), and (7)

$$G(w, x_b) = B[p - q_b(x_b)]^{1/a} w^{-(1-\alpha)/\alpha} - Eu^{-1/a}[w - q_c(x_b)]^{1/a}$$
$$\times \left[Y_0 + Cw^{-(1-\alpha)/\alpha} \int_0^{x_b} [p - q_b(x)]^{(1-\alpha)/\alpha} dx \right]^{-b/a} t_0^{c/a}. \quad (23)$$

We can readily see that the curve representing $G(w, x_b) = 0$ is continuous for $0 < x_b < \bar{x}$ and $w > q_c(0)$, since

$$(dw/dx_b)|_{G=0} = -(\partial G/\partial x_b)/(\partial G/\partial w) \quad (24)$$

and $\partial G/\partial w \neq 0$. Furthermore, when $x_b = 0$, we have $G(w, 0) = 0$, yielding

$$w^{(1-\alpha)/\alpha}[w - q_c(0)]^{1/a} = BE^{-1}u^{1/a}[p - q_b(0)]^{1/\alpha}Y_0^{b/a}t_0^{-c/a} > 0. \quad (25)$$

This implies that

$$w > q_c(0) \quad \text{as} \quad x_b \to 0. \quad (26)$$

On the other hand, when $x_b = \bar{x}$, we have $G(w, \bar{x}) = 0$, giving

$$Eu^{-1/a}[w - q_c(\bar{x})]^{1/a}$$
$$\times \left[Y_0 + Cw^{-(1-\alpha)/\alpha} \int_0^{\bar{x}} [p - q_b(x)]^{(1-\alpha)/\alpha} dx \right]^{-b/a} t_0^{c/a} = 0; \quad (27)$$

i.e.,

$$w \to q_c(x) \quad \text{as} \quad x_b \to \bar{x}. \quad (28)$$

Thus, in the (x_b, w) plane, the curve representing $G(w, x_b) = 0$ is located

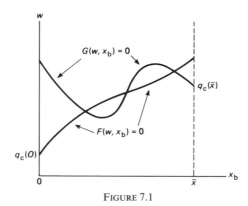

FIGURE 7.1

above the curve representing $F(w, x_b) = 0$ when $x_b = 0$, and the former is below the latter when $x_b = \bar{x}$. As a result, the two curves must intersect at least once, as illustrated in Fig. 7.1. This proves the existence of a set of values, $w > q_c(0)$ and $0 < x_b$.

Finally, condition (11) will be ensured by assuming that the manufacturing bid rent curve is no less elastic than the residential bid rent curve at any location x,

$$-[r_b'(x)/x][x/r_b(x)] \geqq -[r_c'(x)/x][x/r_c(x)] \qquad \text{for all} \quad x. \qquad (29)$$

In view of Eqs. (2) and (7), condition (29) is equivalent to

$$\frac{1}{\alpha}\frac{q_b'(x)}{p - q_b(x)} \geqq \frac{1}{a}\frac{q_c'(x)}{w - q_c(x)} - \frac{c}{a}\frac{1}{t_0 + x - x_b} \qquad \text{for} \quad 0 \leqq x \leqq x_c,$$

or $\qquad\qquad\qquad\qquad\qquad\qquad\qquad\qquad\qquad\qquad\qquad\qquad\qquad\qquad\qquad$ (30)

$$\frac{q_b'(x)}{r_b(x)h_b(x)} \geqq \frac{q_c'(x)}{r_c(x)h_c(x)} - \frac{c}{a}\frac{1}{t_0 + x - x_b} \qquad \text{for} \quad 0 \leqq x \leqq x_c,$$

where $h_b(x) \equiv T(x)/Y(x)$, i.e., land per unit of output. Obviously, this is a modified version of von Thünen's condition, stating that the ratio of marginal transport cost to land rent for the manufacturing sector should be no less than that for the residential sector minus some positive number which is associated with pollution externalities, where we implicitly assume that $t_0 + x - x_b > 0$ for $0 \leqq x \leqq x_c$ so that the last term in condition (30) is always positive. Note that condition (30) is more likely to be satisfied with a higher value of c relative to a, i.e., a higher value of the elasticity of pollution with respect to distance relative to the elasticity of utility with respect to land.

In summary, we have proved the following existence theorem.

Theorem 7.1   *In the present Cobb–Douglas case with zero opportunity cost of land, there exists a spatial equilibrium, i.e., a set of values $w > q_c(0)$, $0 < x_b < \bar{x}$, and $x_c > x_b$, satisfying Eqs. (1)–(11), if conditions (12), (20), and (30) are met.*

7.3   Uniqueness and Stability

We are now in a position to examine the uniqueness property of the equilibrium. It turns out that the assumptions which have been made so far are not sufficient to ensure the uniqueness of the equilibrium, and we need an additional assumption whose validity depends on the degree of external diseconomy.

First, it is straightforward to see from Eq. (23) that

$$\partial G/\partial w < 0. \tag{31}$$

However, we have

$$\frac{\partial G}{\partial x_b} = -\frac{r_b(x_b)}{\alpha}\frac{q_b'(x_b)}{p - q_b(x_b)} + \frac{r_c(x_b)}{a}\frac{q_c'(x_b)}{w - q_c(x_b)}$$

$$+ \frac{b}{a}\frac{r_c(x_b)[p - q_b(x_b)]^{(1-\alpha)/\alpha}}{Y_0 + Y} \gtreqless 0,$$

according as

$$\frac{q_b'(x_b)}{h_b(x_b)} \gtreqless \frac{q_c'(x_b)}{h_c(x_b)} + \frac{b}{a}\frac{r_b(x_b)[p - q_b(x_b)]^{(1-\alpha)/\alpha}}{Y_0 + Y}, \tag{32}$$

where we note that $r_b(x_b) = r_c(x_b)$. If the left-hand side is no less than the right-hand side in condition (32), then

$$\partial G/\partial x_b \leq 0, \tag{33}$$

and therefore, in view of (31),

$$(dw/dx_b)|_{G=0} = -(\partial G/\partial x_b)/(\partial G/\partial w) \leq 0, \tag{34}$$

meaning that in the (x_b, w) plane the curve representing $G(w, x_b) = 0$ has a nonpositive slope everywhere. This ensures the uniqueness of the equilibrium, as seen in Fig. 7.2.

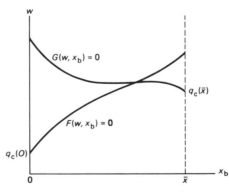

FIGURE 7.2

Theorem 7.2 *The equilibrium whose existence is ensured in Theorem 7.1 is unique, if*

$$\frac{q_b'(x_b)}{h_b(x_b)} \geq \frac{q_c'(x_b)}{h_c(x_b)} + \frac{b}{a} \frac{r_b(x_b)[p - q_b(x_b)]^{(1-\alpha)/\alpha}}{Y_0 + Y} \tag{35}$$

for all w and x_b such that $w > q_c(0)$ and $0 < x_b < \bar{x}$.

It should be noted that condition (35) is somewhat stronger than the usual von Thünen condition and is more likely to be met with a smaller value of b relative to a, i.e., a smaller value of the elasticity of utility with respect to pollution relative to the elasticity of utility with respect to residential land. In fact, if there is no pollution externality so that $b = c = 0$, then condition (30) implies condition (35); i.e., no additional assumption is needed to obtain uniqueness.

Next, let us investigate the stability property of the equilibrium by assuming a dynamic adjustment process of the wage rate and the CBD boundary. It turns out that we need exactly the same additional assumption as in Theorem 7.2 in order to ensure the dynamic stability of the equilibrium for all positive speeds of adjustment. In particular, we assume that the wage rate is adjusted in response to the excess demand for labor as[3]

$$\dot{w} = f[F(w, x_b)] \tag{36}$$

with $f'(\ ) > 0$, $f(0) = 0$, where the dot denotes differentiation with respect to time. On the other hand, the CBD boundary is adjusted in response to the rent differential at that boundary as

$$\dot{x}_b = g[G(w, x_b)] \tag{37}$$

with $g'(\ ) > 0$, $g(0) = 0$. The urban boundary x_c is, however, assumed to be instantaneously adjusted so as to yield $r_c(x_c) = 0$.

Here we shall focus on local stability in a small neighborhood of equilibrium by linearizing Eqs. (36) and (37) around the equilibrium values of w and x_b, namely, w^* and x_b^*;

$$\begin{bmatrix} \dot{w} - w^* \\ \dot{x}_b - x_b^* \end{bmatrix} = \begin{bmatrix} f'(0) \, \partial F/\partial w & f'(0) \, \partial F/\partial x_b \\ g'(0) \, \partial G/\partial w & g'(0) \, \partial G/\partial x_b \end{bmatrix} \begin{bmatrix} w - w^* \\ x_b - x_b^* \end{bmatrix}. \tag{38}$$

Considering properties (17), (31), and (33), we can show that the equilibrium is locally stable for any positive speeds of adjustment $f'(0) > 0$ and $g'(0) > 0$,

---

[3] For a detailed discussion of the adjustment process of this kind, see Chapter 1.

if condition (35) is satisfied, because in that case, the trace of the matrix in system (38) is negative,

$$f'(0) \, \partial F/\partial w + g'(0) \, \partial G/\partial x_b < 0, \tag{39}$$

and the determinant is positive,

$$f'(0)g'(0)[(\partial F/\partial w)(\partial G/\partial x_b) - (\partial G/\partial w)(\partial F/\partial x_b)] > 0. \tag{40}$$

Theorem 7.3 *The equilibrium whose existence is proved in Theorem 7.1 is locally stable for any positive speeds of adjustment $f'(0) > 0$ and $g'(0) > 0$, if condition (35) is satisfied at the equilibrium.*

Finally it should be noted that if the degree of external diseconomy b is so large as to give

$$\frac{q_b'(x_b)}{h_b(x_b)} < \frac{q_c'(x_b)}{h_c(x_b)} + \frac{b}{a} \frac{r_b(x_b)[p - q_b(x_b)]^{(1-\alpha)/\alpha}}{Y_0 + Y} \tag{41}$$

for all possible values of w and x_b, then there will exist a set of positive speeds of adjustment $f'(0)$ and $g'(0)$ (the latter is sufficiently larger than the former) such that the trace becomes positive, i.e.,

$$f'(0) \, \partial F/\partial w + g'(0) \, \partial G/\partial x_b > 0$$

for all possible equilibria. This means the following proposition.

Theorem 7.4 *If condition (41) holds for all possible values of w and x_b, then there exist a set of positive speeds of adjustment for which none of the equilibria will be locally stable.*

The results obtained above have particular significance for applied work on the basis of urban location theory with pollution externalities. Especially, important implications can be derived regarding the so-called hedonic price approach to evaluating environmental quality.[4] It has been shown by Polinsky and Shavell (1976) that hedonic prices should properly be interpreted only within the context of spatial equilibrium theory. The disturbing feature of our results is the implication that the hedonic price may not be unique and may not even be observed, as the uniqueness and stability of the underlying spatial equilibrium may not be ensured without some prior restrictions on the degree of external diseconomies. Although this fact does not deny the applicability of the hedonic price approach, one must be careful in using it for making a policy argument in the context of spatial equilibrium with external diseconomies.

---

[4] Typical examples of the hedonic price approach are those of Mieszkowski and Saper (1978) and Nelson (1978).

REFERENCES

Henderson, J. V. (1977a). Externalities in a spatial context. *Journal of Public Economics* **7**, 89–110.

Henderson, J. V. (1977b). *Economic Theory and the Cities*. Academic Press, New York.

Hochman, O. (1978). A two sector model with several externalities and their effects on an urban setting. *Journal of Urban Economics* **5**, 198–218.

Mieszkowski, P., and Saper, A. M. (1978). An estimate of the Effects of airport noise on property values. *Journal of Urban Economics* **5**, 425–440.

Miyao, T., Shapiro, P., and Knapp, D. (1980). On the existence, uniqueness and stability of spatial equilibrium in an open city with externalities. *Journal of Urban Economics* **8**, 139–149.

Nelson, J. P. (1978). Residential choice, hedonic prices and the demand for urban air quality. *Journal of Urban Economics* **5**, 357–369.

Polinsky, A. M., and Shavell, S. (1976). Amenities and property values in a model of an urban area. *Journal of Public Economics* **5**, 119–129.

Starrett, D. A. (1972). Fundamental nonconvexities in the theory of externalities. *Journal of Economic Theory* **4**, 180–199.

Stull, W. J. (1974). Land use and zoning in an urban economy. *American Economic Review* **64**, 337–347.

PART 3

DYNAMICS OF URBAN GROWTH

CHAPTER 8

Industrial Growth

No one fails to notice that there have been an increasing number of economic and social problems arising from competition among growing industries over available land and from conflicts between rapid economic growth and optimal land use in and around urban areas.[1] In the literature, however, very little work has been done to formulate appropriate urban growth models which incorporate both spatial elements and growth factors explicitly to study the phenomenon of long run urban growth over time *and space*.[2] In this chapter we take a step toward a full integration of urban location theory with economic growth theory by constructing a simple industrial growth model with spatial elements. For the sake of illustration, we shall focus on only two factors which seem to be among the most important ones for long-run industrial growth over space, namely, population growth and transportation improvements.

First, we deal with the type of industrial growth caused by exogenous population growth accompanied by endogenous transportation improvements. The model assumes that population is growing at a given constant rate while transportation improvements are brought about by social transportation investment which is financed by taxation. Then we carry out a

---

[1] This chapter is an extended version of Miyao (1977).

[2] Quite recently, several authors have developed urban growth models with particular emphasis on durable housing; e.g., Anas (1978), Arnott (1980), Brueckner (1980), Fujita (1976), and Hochman and Pines (1980). On the other hand, our focus here is on long-run dynamics where structural durability can be ignored.

dynamic analysis of urban growth by identifying a balanced growth equilibrium and proving its existence, uniqueness, and global stability under some assumptions.

From our model, we can answer the following often-asked question: How much, or how much more, should we invest for transportation improvements, e.g., new highway construction, in an ever growing city? It seems that while more and more resources will be required to improve transportation facilities as city size increases, total output and income for the city as a whole will also grow as more capital and labor become available in the city. Therefore, a proper way of posing the question in the context of urban growth is to ask what fraction of total output (income) should be devoted to transportation investment in order to maximize net output per capita in the city. In our model, we find a very simple answer to this question, i.e., an optimal taxation rule for transportation improvements, which is analogous to the well-known golden rule of capital accumulation.

We also consider an alternative model of urban growth, in which transportation improves exogenously with no investment required in the city, whereas population grows due to net in-migration, the rate of which depends on the difference between the wage rate in the city and the national wage rate. The existence, uniqueness, and global stability of a balanced growth equilibrium will be examined in this model.

8.1   Exogenous Population Growth

In this section, we shall set up an urban growth model with population growing exogenously and transportation improving endogenously. First, we study an industrial location model which is the simplest case of the general model developed in Chapter 1. There is assumed to be a circular city filled with a single group of identical firms producing the same commodity subject to identical production functions, using land and labor. All products must be transported to the central market of the city with the road system being radial and dense. Transport cost per unit of product is a sole function of distance and, more specifically, is *linearly* related to distance. It is also assumed that the city is a small open economy so that product price is given exogenously.

Since production cost per unit of output, C, at distance x is a function of land rent $r(x)$ at x and the wage rate w, the zero profit condition under perfect competition is written as

$$C[r(x), w] = 1 - qx, \tag{1}$$

where the product price is normalized as unity and q is the marginal (unit) transport cost which is constant by assumption. It is a well-known property of the unit cost function C that the labor–land ratio $n(x)$ at x can be derived as

$$n(x) = (\partial C/\partial r)/(\partial C/\partial w) \equiv n[r(x), w]. \tag{2}$$

Assume that a constant fraction g of land at each distance x is used for production, where $0 < g \leq 1$. Then the land available for production in the ring with inner radius x and outer radius $x + dx$ is $2\pi g x\, dx$, and the amount of labor employed in that ring is $2\pi g n(x) x\, dx$. The full employment condition is thus

$$2\pi g \int_0^{x_b} n(x)x\, dx = N, \tag{3}$$

where x_b is the distance from the city center to the "urban" boundary and N is the total amount of labor available in the city. The location of the urban boundary x_b is determined by the market rent condition that the overall rent function should be continuous at x_b;

$$r(x_b) = r_0, \tag{4}$$

where r_0 is the opportunity cost of land, e.g., agricultural land rent. The system (1)–(4) can determine the equilibrium functional forms of $r(x)$ and $n(x)$ and the equilibrium values of w and x_b simultaneously, given q, N, and r_0.

Now we introduce dynamic changes in unit transport cost q and total labor force N. As in the ordinary neoclassical growth theory, population growth is considered to play a basic role for propelling urban growth in the long run and thus treated as exogenous in our model,

$$\hat{N} = \mu > 0, \tag{5}$$

where the hat denotes logarithmic differentiation with respect to time, and μ is a given positive constant. On the other hand, transportation improvements as represented by reductions in unit transport cost should be regarded as endogenous, since it is obvious that any major transportation improvement, e.g., new highway construction, requires an enormous amount of resources which could be used to meet various other needs of the city. Specifically, we assume that the rate of decrease of unit transport cost depends on the amount of transportation investment per unit of land and that the total amount of transportation investment S is evenly distributed all over the land area devoted to transportation, J, so as to produce a uniform rate of reduction of unit transport cost everywhere in the city,

$$-\hat{q} = \lambda(S/J), \tag{6}$$

with[3] $\lambda'(\ ) > 0$, $\lambda(0) = 0$, $\lambda(\infty) = \infty$, where

$$J \equiv 2\pi c \int_0^{x_b} x \, dx = \pi c(x_b)^2, \tag{7}$$

and c is the constant fraction of land available for transportation.

Let us assume that transportation investment is financed by a proportional sales tax, or equivalently a proportional excise tax, levied at the central market. Then the fact that total transportation investment equals total tax revenue can be written as

$$S = 2\pi g\tau \int_0^{x_b} f[n(x)]x \, dx, \tag{8}$$

where τ is the tax rate per unit of output and $f[n(x)]$ is the output–land ratio expressed as a function of the labor–land ratio $n(x)$ at x. Under the present assumption on taxation, Eq. (1) becomes

$$C[r(x), w] = 1 - \tau - qx, \tag{9}$$

which may be solved for r as

$$r(x) = r(1 - \tau - qx, w). \tag{10}$$

In view of Eqs. (9) and (10), we derive $n(x) = n[r(1 - \tau - qx, w), w]$ from Eq. (2) and $x_b = [1 - \tau - C(r_0, w)]/q$ from Eq. (4). Then, letting $z \equiv qx$, we can rewrite Eq. (3) as

$$G(w) \equiv 2\pi g \int_0^{1-\tau-C(r_0,w)} n[r(1 - \tau - z, w), w]z \, dz = Nq^2, \tag{11}$$

and we can easily see that $G'(w) < 0$, since $\partial r/\partial w < 0$, $\partial n/\partial r > 0$, $\partial n/\partial w < 0$, and $\partial C/\partial w > 0$. If we assume[4]

$$n[r(1 - \tau - z, 0), 0] = \infty, \qquad n[r(1 - \tau - z, \infty), \infty] = 0,$$
$$\text{for} \quad 0 \leq z \leq 1 - \tau, \tag{12}$$

and[5]

$$1 - \tau - C(r_0, \infty) > 0, \tag{13}$$

---

[3] These restrictions can be relaxed as follows. There exist nonnegative constants $\underline{z}$ and $\bar{z}$ such that $\lambda(z) < \mu/2$ for $z \leq \underline{z}$, $\lambda'(z) > 0$ for $\underline{z} < z < \bar{z}$, and $\lambda(z) > \mu/2$ for $z \geq \bar{z}$. This will allow transportation *disimprovements*, i.e., $\lambda(0) < 0$, in the absence of transportation investment due to increasing energy costs, for example.

[4] This is, in fact, satisfied in the Cobb–Douglas case.

[5] This condition is met in the Cobb–Douglas case with zero-opportunity cost of land, as we will consider later.

then we have $G(0) = \infty$ and $G(\infty) = 0$ and, therefore, defining

$$k \equiv 1/(Nq^2), \tag{14}$$

we can solve Eq. (11) for w as $w = w(k) > 0$ for $0 < k < \infty$, and obviously $w'(k) > 0$ for all k.

Now our dynamic system may be described in terms of k. In fact, it follows from Eqs. (5), (6), and (14) that

$$\hat{k} = -2\hat{q} - \hat{N} = 2\lambda(S/J) - \mu = H(k) - \mu, \tag{15}$$

where

$$H(k) \equiv 2\lambda \left[\frac{2g\tau}{c\{1 - \tau - C[r_0, w(k)]\}^2} \right.$$
$$\left. \times \int_0^{1-\tau-C[r_0,w(k)]} f(n\{r[1 - \tau - z, w(k)], w(k)\})z \, dz \right]. \tag{16}$$

It is clear from Eq. (15) that a balanced growth equilibrium can be defined as a state in which k is stationary over time, i.e., $0 = \hat{k} = -2\hat{q} - \mu$, or equivalently, w is stationary over time. This means that in a balanced growth equilibrium the rate of decrease of unit transport cost is just half the rate of population growth, and that the ratio of total labor force N to total city area $\pi(x_b)^2$ remains stationary along a balanced growth equilibrium path, because

$$\widehat{(x_b)^2} = 2\hat{x}_b = -2\hat{q} = \mu$$

with a stationary value of w.

We can prove the existence of a balanced growth equilibrium and the global stability of the dynamic system, if we further assume[6]

$$f(0) = 0, \qquad f(\infty) = \infty. \tag{17}$$

In view of Eqs. (11) and (13), we find

$$n \to \infty \quad \text{as} \quad k \to 0 \;\; (Nq^2 \to \infty), \qquad n \to 0 \quad \text{as} \quad k \to \infty \;\; (Nq^2 \to 0).$$

Then it follows from Eqs. (13), (16), and (17) that $H(0) = \infty$, $H(\infty) = 0$. By virtue of the continuity of $H(k)$, there must exist a value of k such that $\hat{k} = H(k) - \mu = 0$. Although uniqueness is not in general ensured, global stability can be obtained. Since $\hat{k} > 0$ as $k \to 0$, and $\hat{k} < 0$ as $k \to \infty$, the dynamic system is globally stable in the sense that, regardless of its initial value, k will asymptotically converge to a stationary value such that $H(k) = \mu$.

---

[6] This is again satisfied in the Cobb–Douglas case.

Theorem 8.1   *In the industrial city with exogenous population growth and a proportional sales (excise) tax to finance transportation improvements, there exists a balanced growth equilibrium, and the dynamic system is globally stable under the present assumptions, especially conditions* (7), (12), (13), *and* (17).

8.2 Optimal Taxation

In order to proceed further, we shall study a special case in which the production function is Cobb–Douglas, $f(n) = n^{1-a}$, where a is the land elasticity of output. The zero profit condition (9) will then be

$$a^{-a}(1 - a)^{-(1-a)}r(x)^a w^{1-a} = 1 - \tau - qx,$$

and therefore Eq. (10) becomes

$$r(x) = a(1 - a)^{(1-a)/a}w^{-(1-a)/a}(1 - \tau - qx)^{1/a}. \tag{18}$$

In this special case, it is easy to see that $n(x) = [(1 - a)/a][r(x)/w]$, which, together with Eq. (18), leads to

$$n(x) = (1 - a)^{1/a}w^{-1/a}(1 - \tau - qx)^{1/a}.$$

For the sake of simplicity, we assume that the opportunity cost of land is negligible around the urban boundary, i.e., $r_0 = 0$, which implies $x_b = (1 - \tau)/q$. Then Eq. (11) becomes

$$2\pi g(1 - a)^{1/a}w^{-1/a}\int_0^{1-\tau}(1 - \tau - z)^{1/a}z\,dz$$

$$= 2\pi g(1 - a)^{1/a}w^{-1/a}\int_0^{1-\tau}z^{1/a}(1 - \tau - z)\,dz = 1/k,$$

or

$$w^{-(1-a)/a} = A(1 - \tau)^{-(1+2a)(1-a)/a}k^{-(1-a)}, \tag{19}$$

where A is a positive constant which does not depend on k or τ. Our dynamic system is expressed as $\hat{k} = H(k) - \mu$, with

$$H(k) = 2\lambda\left[2gc^{-1}\tau(1 - \tau)^{-2}(1 - a)^{(1-a)/a}w^{-(1-a)/a}\right.$$

$$\left.\times \int_0^{1-\tau}(1 - \tau - z)^{(1-a)/a}z\,dz\right]$$

$$= 2\lambda[B\tau(1 - \tau)^{-2(1-a)}k^{-(1-a)}], \tag{20}$$

where B is a positive constant. Thus we can easily prove the existence, uniqueness, and global stability of a balanced growth equilibrium just as done by Solow (1956), since it is clear from equation (20) that

$$H(0) = \infty, \qquad H(\infty) = 0, \qquad \text{and} \qquad H'(k) < 0 \qquad \text{for all} \quad k > 0.$$

Theorem 8.2 *A balanced growth equilibrium exists and is unique and globally stable in the Cobb–Douglas case with zero-opportunity cost of land, where population grows exogenously and transportation improvements are financed by a proportional sales tax.*

Next, let us focus on the state of balanced growth equilibrium and determine the optimal tax rate so as to maximize the balanced growth equilibrium level of "net" output per capita $y \equiv Y/N$, where total net output Y is defined as total gross output minus total tax payment and total transport cost.

$$Y \equiv 2\pi g \int_0^{x_b} (1 - \tau - qx) f[n(x)] x \, dx.$$

In the Cobb–Douglas case, we have

$$y = Y/N = D(1 - \tau)^{(1 + 2a)/a} w^{-(1-a)/a} k = E(1 - \tau)^{(1+2a)} k^a,$$

where D and E are some positive constants. Then the problem considered here is to maximize y with respect to τ subject to $0 = \hat{k} = H(k) - \mu$, or, equivalently, to maximize $(1 - \tau)^{(1+2a)} k^a$ with respect to τ subject to

$$2\lambda [B\tau(1 - \tau)^{-2(1-a)} k^{-(1-a)}] = \mu.$$

This reduces to maximizing $\tau^{a/(1-a)}(1 - \tau)$ with respect to τ, and from this it follows that y is maximized if and only if

$$\tau = a, \tag{21}$$

i.e., the optimal tax rate is equal to the elasticity of output with respect to land. Furthermore, in the Cobb–Douglas case, we know that

$$a = r(x)/\{(1 - \tau - qx) f[n(x)]\} \qquad \text{for all} \quad x,$$

which means that the land elasticity of output a equals the ratio of total land rent to total net output Y.

Thus we have shown the following proposition.

Theorem 8.3 *In the Cobb–Douglas case with no-opportunity cost of land, the balanced growth equilibrium level of net output per capita is maximized, if and only if the rate of the proportional sales tax to finance transportation* *improvements is equal to the land elasticity of output, i.e., the land share in net output.*

Obviously, this result is similar to the well-known golden rule that the optimal saving ratio is equal to the capital elasticity of output in the ordinary neoclassical growth model (Phelps, 1961). In fact, in our model, taxation is the only form of savings from the social point of view, and transportation investment is the only way to reduce unit transport cost and thereby increase the total amount of land used in production, just as capital investment is to increase the total amount of capital in the ordinary growth model.

8.3   More on the Golden Rule

The similarity of our result to the standard golden rule result can be seen more clearly, if we assume either that transport cost is incurred in the form of product depreciation or that transport cost is tax deductible. In either case, the zero profit condition may be written as

$$a^{-a}(1 - a)^{-(1-a)}r(x)^a w^{1-a} = (1 - qx)(1 - \tau), \tag{22}$$

and the land–labor ratio as

$$n(x) = w^{-1/a}(1 - a)^{1/a}(1 - \tau)^{1/a}(1 - qx)^{1/a}. \tag{23}$$

The equality of total transportation investment and total tax revenue is

$$S = \tau Q, \tag{24}$$

where Q is defined as total output which equals total gross output minus total transport cost,

$$Q = 2\pi g \int_0^{x_b} (1 - qx)n(x)^{1-a}x \, dx$$

$$= 2\pi g w^{-(1-a)/a}(1 - a)^{(1-a)/a}(1 - \tau)^{(1-a)/a} \int_0^{x_b} (1 - qx)^{1/a}x \, dx. \tag{25}$$

On the other hand, the full employment condition (3) becomes

$$2\pi g w^{-1/a}(1 - a)^{1/a}(1 - \tau)^{1/a} \int_0^{x_b} (1 - qx)^{1/a}x \, dx = N, \tag{26}$$

which, together with Eq. (25), implies

$$Q/N = w/[(1 - a)(1 - \tau)]. \tag{27}$$

It is convenient to define

$$L \equiv 2\pi g \int_0^{x_b} (1 - qx)^{1/a} x \, dx, \tag{28}$$

which might be called the "efficient stock of land," consisting of all individual pieces of land in the city with the respective weights of "efficiency" $(1 - qx)^{1/a}$ associated with distance x from the central market. Here, any difference in efficiency between two locations is due solely to the transport cost difference between the two locations in question. Obviously, the efficient stock of land is a similar concept to the efficient stock of capital used in the context of economic growth and technical progress (Solow, 1960). Utilizing the concept (28), we find from Eq. (26) that

$$(L/N)^a = w/[(1 - a)(1 - \tau)],$$

which, together with Eq. (27), will yield the "aggregate production function"

$$Q/N = (L/N)^a \quad \text{or} \quad Q = L^a N^{1-a}. \tag{29}$$

In view of the fact that $x_b = 1/q$ from Eq. (22) with $r(x_b) = 0$ and $0 < \tau < 1$, we also find

$$L = 2\pi g q^{-2} \int_0^1 (1 - z)^{1/a} z \, dz = M/q^2, \tag{30}$$

where M is a positive constant. It follows from Eqs. (14), (29), and (30) that $Q/N = M^a k^a$. In the present case, therefore, the dynamic equation is written as

$$\hat{k} = 2\lambda(\tau Q/[\pi c/q^2]) - \mu = 2\lambda(\tau M^a k^a/[\pi c k]) - \mu = G(\tau k^a/k) - \mu, \tag{31}$$

where it is obvious that the function G has exactly the same properties as the function H in Eq. (20), so that the existence, uniqueness, and global stability of a balanced growth equilibrium can be proved just as before.

Our optimization problem now is to maximize the balanced growth equilibrium level of per capita output net of tax,

$$(1 - \tau)Q/N = (1 - \tau)M^a k^a,$$

subject to $0 = \hat{k} = G(\tau k^a/k) - \mu$ or, equivalently,

$$\text{maximize} \quad (1 - \tau)k^a \quad \text{subject to} \quad \tau k^a/k = G^{-1}(\mu). \tag{32}$$

As can be easily noticed, this problem is formally equivalent to the standard golden rule problem (Phelps, 1961), in which the saving ratio s and the natural rate of growth μ are substituted for τ and $G^{-1}(\mu)$, respectively, in problem (32). Thus, the answer is again $\tau = a$. It may be useful to look at

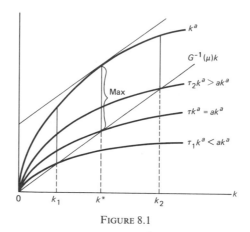

FIGURE 8.1

Fig. 8.1, noting that $dk^a/dk = ak^{a-1}$ and at the optimum we have $a(k^*)^{a-1} = G^{-1}(\mu)$, or $a(k^*)^a = G^{-1}(\mu)k^*$, which leads to the golden rule $\tau = a$.

In summary, we have proved the following theorem.

Theorem 8.4 *If transport cost takes a form of product depreciation or is tax deductible in the Cobb–Douglas case with no-opportunity cost of land and a proportional sales tax to finance transportation improvements, then* (i) *a balanced growth equilibrium exists and is unique and globally stable for each given tax rate, and* (ii) *the equilibrium level of per capita output net of tax is maximized when the tax rate equals the land elasticity of output, i.e., the land share in net output.*

8.4 Exogenous Transportation Improvements

So far we have treated population growth as exogenous and transportation improvements as endogenous in the model. In some cases, particularly in the case of relatively small open cities, however, it might be more appropriate to assume that transportation improves exogenously due to the spillover (free-rider) effects of transportation improvements taking place in larger or nearby jurisdictions,[7] whereas population growth is affected by in- and out-migration and, therefore, is treated as endogenous.

---

[7] This also could be due to transportation investment being financed by interjurisdictional transfers, e.g., federal grants.

More specifically, unit transport cost is assumed to be decreasing at a constant exogenous rate

$$-\hat{q} = \lambda \geq 0, \tag{33}$$

with no investment (and no tax) required in the city. On the other hand, the rate of net inflow of population to the city depends on the difference between the wage rate in this city and the national wage rate which is given to the city,

$$\hat{N} = \mu(w - \bar{w}), \tag{34}$$

with

$$\mu'(\ ) > 0, \qquad \mu(0) = 0, \qquad \mu(\infty) = \infty. \tag{35}$$

Setting $\tau = 0$ in Eq. (11) and noting $k = 1/(Nq^2)$, we find

$$G(w) = 1/k \qquad \text{or} \qquad w = G^{-1}(1/k). \tag{36}$$

Then, the dynamic equation (15) becomes

$$\hat{k} = -2\hat{q} - \hat{N} = 2\lambda - \mu\lfloor G^{-1}(1/k) - \bar{w}\rfloor. \tag{37}$$

A balanced growth equilibrium is obtained by setting $\hat{k} = 0$. In view of the fact that

$$d\hat{k}/dk = -\mu' G^{-1\prime}(-1/k^2) < 0 \qquad \text{for all} \quad k > 0, \tag{38}$$

and also the properties of $\mu(\ )$ and $G^{-1}(\ )$, we can show the existence, uniqueness, and global stability of the equilibrium, as in Fig. 8.2.

In economic terms, the stability property follows from direct responses of

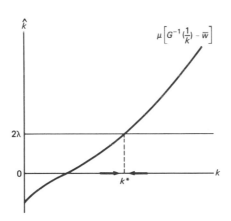

FIGURE 8.2

population to changes in the wage rate. If the initial value of k is greater (resp. smaller) than the equilibrium value k^*, i.e., if the initial size of population is smaller (resp. larger) than its equilibrium value in relation to unit transport cost, then the marginal product of labor and therefore the wage rate will be greater (resp. smaller) than their respective equilibrium values, and thus population will be growing at a faster (resp. slower) rate than two times the rate of decrease of unit transport cost. As a result, k will be decreasing (resp. increasing) toward k^*.

Theorem 8.5   *In the industrial city with exogenous transportation improvements and endogenous population growth due to net in-migration, a balanced growth equilibrium exists and is unique and globally stable under the assumptions made in Sections 8.1 and 8.4, particularly conditions (12), (13), (17), and (35).*

REFERENCES

Anas, A. (1978). Dynamics of urban residential growth. *Journal of Urban Economics* **5,** 66–87.
Arnott, R. J. (1980). A simple urban growth model with durable housing. *Regional Science and Urban Economics* **10,** 53–76.
Brueckner, J. K. (1980). Residential succession and land use dynamics in a vintage model of urban housing. *Regional Science and Urban Economics* **10,** 225–240.
Fujita, M. (1976). Spatial patterns of urban growth: Optimum and market. *Journal of Urban Economics* **3,** 209–241.
Hochman, O., and Pines, D. (1980). Costs of adjustment and demolition costs in residential construction and their effects on urban growth. *Journal of Urban Economics* **7,** 2–19.
Miyao, T. (1977). A long-run analysis of urban growth over space. *Canadian Journal of Economics* **10,** 678–686.
Phelps, E. S. (1961). The golden rule of accumulation: A fable for growthmen. *American Economic Review* **51,** 638–643.
Solow, R. M. (1956). A contribution to the theory of economic growth. *Quarterly Journal of Economics* **70,** 65–94.
Solow, R. M. (1960). Investment and technical progress. In *Mathematical Methods in the Social Sciences* (K. J. Arrow, S. Karlin, and P. Suppes, eds.), pp. 89–104. Stanford Univ. Press, Stanford, California.

CHAPTER 9

Residential Growth

In this chapter we turn our attention to a residential city which is expanding over time and space.[1] First, we consider a residential location model with a single class of identical households and dynamize the model by introducing exogenous population growth and endogenous transportation improvements. Assuming that transportation investment is financed by a proportional income tax, we prove the existence, uniqueness, and global stability of a balanced growth equilibrium for each given tax rate. Then we find the optimal tax rate so as to maximize the long-run equilibrium level of utility for all households in the city. As in the previous chapter, our result on optimal taxation for transportation improvements is quite analogous to the well-known golden rule of capital accumulation.

It is also shown that our simple golden rule remains valid in the case of two residential classes with different incomes, so long as utility functions are all identical and Cobb–Douglas. Finally, some modifications are made concerning the opportunity cost of land and the rate of discount for future consumption, and a relation between our result and the so-called "turnpike theorem" is pointed out.

9.1 The Model with One Residential Class

As in Chapter 2, we assume a monocentric city where every resident commutes to the CBD. Transport (commuting) cost is assumed to be a linear

---

[1] This chapter is largely based on Miyao (1977).

function of distance. For the moment, we suppose that there is only one class of identical residents having identical incomes and identical utility functions.

Let us take a resident living at distance x from the CBD. His utility is assumed to depend on the amount of the consumption good d consumed and the residential space h occupied at x:

$$U(x) = U[d(x), h(x)]. \tag{1}$$

His budget constraint can be written as

$$d(x) + r(x)h(x) = 1 - qx, \tag{2}$$

where $r(x)$ is the residential land rent at x, q the marginal transport cost per person which is a given constant, and his income and the price of the consumption good are both given and normalized as unity.[2]

Maximization of the utility function (1) subject to the budget constraint (2) leads to

$$u = V[r(x), 1 - qx], \tag{3}$$

where V is the indirect utility function, depending on the relative price $r(x)$ and net income $1 - qx$, and u is the maximized level of utility which is independent of x, because every resident should attain the same level of utility in equilibrium in the case of identical residents. From Eq. (3), we can derive the rent function $r(x)$ as

$$r(x) = r(1 - qx, u), \tag{4}$$

and the demand for land per person at x as

$$h(x) = -(\partial V/\partial r)/[\partial V/\partial(1 - qx)] \equiv h[r(x), 1 - qx]. \tag{5}$$

The full accommodation condition that all residents should be housed within the city can be expressed as

$$2\pi g \int_0^{xc} \frac{x}{h[r(x), 1 - qx]} dx = N, \tag{6}$$

where x_c is the distance from the CBD to the urban boundary, N the total number of residents in the city, and g the constant fraction of land devoted to residential use at each distance from the CBD. Finally, in equilibrium the overall rent function should be continuous at the urban boundary,

$$r(x_c) = r_0, \tag{7}$$

[2] The price and income can be normalized as unity at the same time by appropriately choosing the units of measurement for the consumption good and labor. It should also be noted that income growth can easily be incorporated into the model without affecting our final results, so long as its growth rate is given exogenously.

where r_0 is the opportunity cost of land, which is exogenously given. Equations (3)–(7) determine the equilibrium functional forms of $r(x)$ and $h(x)$ and the equilibrium values of u and x_c for given values of q, N, and r_0.

Now let us dynamize the model by introducing dynamic changes in unit transport cost q and total population N over time. For our analysis, the best way to proceed is to treat population growth as exogenous and transportation improvements as endogenous. Specifically, we assume that the growth rate of population is exogenously given as a positive constant,

$$\hat{N} = \mu > 0, \tag{8}$$

where the hat denotes logarithmic differentiation with respect to time, and that the rate of reduction in unit transport cost depends on the amount of transportation investment per unit of land devoted to transportation,

$$-q = \lambda(S/J), \tag{9}$$

with[3]

$$\lambda'(\ ) > 0, \qquad \lambda(0) = 0, \qquad \lambda(\infty) = \infty, \tag{10}$$

where S is total transportation investment and J the total land area used for transportation, i.e.,

$$J \equiv 2\pi e \int_0^{x_c} x\, dx = \pi e(x_c)^2, \tag{11}$$

with e the constant fraction of land devoted to transportation. All this amounts to assuming that the total amount of transportation investment is uniformly distributed over the land area used for transportation in the city and that greater reductions in unit transport cost will result everywhere in the city with more investment per unit of land for transportation.

Let us ask how much should be spent for transportation improvements to maximize the long-run utility level of all residents in the city. In order to determine the optimal ratio of transportation investment to income, we shall assume that transportation investment is financed by a proportional income tax, which may also be interpreted as a lump-sum tax in the case of identical residents. In the present case, we have

$$S = \tau N, \qquad 0 < \tau < 1, \tag{12}$$

where the tax rate τ is in fact equal to the amount of tax paid by each resident whose income is normalized as unity. Then the budget constraint (2) becomes

$$d(x) + r(x)h(x) = 1 - \tau - qx. \tag{13}$$

[3] These restrictions may be relaxed as follows. There exist nonnegative constants $\underline{z}$ and $\bar{z}$ such that $\lambda(z) < \mu/2$ for $z \leq \underline{z}$, $\lambda'(z) > 0$ for $\underline{z} < z < \bar{z}$, and $\lambda(z) > \mu/2$ for $z \geq \bar{z}$.

To proceed further, we shall assume that the utility function is Cobb–Douglas,

$$U(x) = d(x)^a h(x)^b, \qquad a > 0, \quad b > 0. \tag{14}$$

Equation (3) will then be

$$u = Ar(x)^{-b}(1 - \tau - qx)^{a+b}, \tag{15}$$

and thus

$$h(x) = B(1 - \tau - qx)^{-a/b} u^{1/b}, \tag{16}$$

where A and B are some positive constants, and Eqs. (6) and (7) become

$$Cu^{-1/b} \int_0^{xc} x(1 - \tau - qx)^{a/b} \, dx = N, \tag{17}$$

and

$$D(1 - \tau - qx_c)^{(a+b)/b} u^{-1/b} = r_0, \tag{18}$$

where C and D are positive constants. With initial values of q and N, the dynamic motion of the model can be completely determined by the equation system (8), (9), (11), (12), and (15)–(18), given τ and r_0.

9.2   Balanced Growth and the Golden Rule

First, we define a balanced growth equilibrium as a state in which the level of utility u is stationary over time. In order to clarify the meaning of this definition, let us set $z \equiv 1 - \tau - qx$ in Eq. (17) and obtain

$$u = \left[\frac{F}{Nq^2} \int_0^{1-\tau} (1 - \tau - z) z^{a/b} \, dz \right]^b = Gk^b (1 - \tau)^{a+2b}, \tag{19}$$

where F and G are some positive constants, and $k = 1/(Nq^2)$. Since $1/q$ represents the distance one can cover with the travel expense equivalent to one unit of the consumption good, k may be interpreted as the area that N residents can cover with the transportation expenditure equal to one unit of the consumption good. In this sense, k can serve as a measure of transportation efficiency in the city as a whole. It follows from Eq. (19) that with a given value of τ, u is stationary if and only if k is stationary over time,

$$0 = \hat{k} = -2\hat{q} - \hat{N}, \tag{20}$$

i.e., in a balanced growth equilibrium we have $-2\hat{q} = \hat{N}$, which together with the property $\hat{x}_c = -\hat{q}$ from Eq. (18) will yield $2\hat{x}_c = \hat{N}$; that is to say, total city area $\pi(x_c)^2$ is growing at the same rate as total population N.

Next, we shall prove the existence, uniqueness, and global stability of a balanced growth equilibrium in terms of the movement of k. Let us, for the moment, assume that there exists plenty of vacant land around the city so that the opportunity cost of land is negligibly small,

$$r_0 = 0. \qquad (21)$$

This assumption is made only for the sake of analytical convenience and will be relaxed later. Under this simplifying assumption, we obtain

$$\hat{k} = 2\lambda[N/\{\pi e(x_c)^2\}] - \mu = 2\lambda[\tau/\{\pi e(1-\tau)^2 k\}] - \mu \qquad (22)$$

from Eqs. (8), (9), (11), (12), (18), and (21). By setting $\hat{k} = 0$, we can determine the balanced growth equilibrium value of k, say k^*, for a given value of τ,

$$k^* = \tau/[\lambda^{-1}(\mu/2)\pi e(1-\tau)^2] > 0 \qquad \text{for} \quad 0 < \tau < 1. \qquad (23)$$

Notice that the existence and uniqueness of k^* are ensured by condition (10). Its global stability follows from the fact that, according to Eq. (22), $\hat{k} > 0$ when k is sufficiently small and $\hat{k} < 0$ when k is sufficiently large. Thus, we have proved the following proposition.

Theorem 9.1 *In the residential city with identical residents having Cobb–Douglas utility functions along with no-opportunity cost of land, a balanced growth equilibrium exists and is unique and globally stable, where population grows exogenously and transportation improves endogenously according to Eqs. (9) and (10).*

In economic terms, the stability property may be explained as follows. Suppose initially transportation is relatively inefficient so that the initial value of k is lower than k^*. This means that the initial city size is small relative to the total number of residents and, therefore, the total tax revenue which depends only on the number of residents is large relative to the city size. As a result, a relatively large amount of money is available for transportation investment per unit of land and transportation will improve quickly relative to population growth. Thus, k will increase and approach k^* in the long run. If, on the other hand, the initial value of k is higher than k^*, the city size is large relative to population, and transportation investment per unit of land is relatively small. Therefore, transportation will improve slowly relative to population, and k will decrease and eventually approach k^*.

Now let us ask what the optimal tax rate is from the long-run point of view, or to put it differently, which balanced growth path ought to be chosen as optimal when different paths correspond to different values of τ according to Eq. (23). Here we need a criterion of optimality in order to answer the question. In the present case of identical residents, we can find a very simple criterion, i.e., the level of utility u, which is the same for all residents in the city. Substitution of the equilibrium value of k [given by Eq. (23)] into Eq. (19) yields the balanced growth equilibrium level of utility, say u^*, as a function of τ,

$$u^* = H\tau^b(1 - \tau)^a, \tag{24}$$

where H is a positive constant. It is easily seen from Eq. (24) that u^* is maximized if and only if

$$\tau = b/(a + b). \tag{25}$$

Here we note that from Eqs. (13) and (14) we have

$$b/(a + b) = r(x)h(x)/(1 - \tau - qx) \qquad \text{for all} \quad x.$$

In summary, we have shown the following.

Theorem 9.2 *In the Cobb–Douglas case with no-opportunity cost of land, the balanced growth equilibrium level of utility is maximized if and only if the (income) tax rate is equal to the ratio of rent payment to net income, i.e., the land share in net income.*

Our result is quite analogous to the Phelps (1961) golden rule result, which states that the optimal saving ratio is equal to the capital share in total income. In fact, in our model, taxation is the only form of savings for the urban economy. On the other hand, land rents arise from locational advantages solely due to positive transportation costs, and land rents are increasing over time owing to cost reductions caused by transportation investment. Thus, total land rent may be viewed as residents' total payment for the accumulated effect of transportation investment or, in other words, for the "transportation capital" embodied in land.

9.3 Two Income Classes

In this section we shall demonstrate that our simple golden rule result remains valid in a more general case with two classes of residents having different income levels, while possessing identical Cobb–Douglas utility functions and identical linear transport cost functions. It can be shown that in equilibrium the lower income class will occupy the inner ring of the city, whereas the higher income class will be located in the outer ring (see, e.g.,

Muth, 1969; Solow, 1973). Let us call the former "class 1" and the latter "class 2."

In the case of identical utility functions, the corresponding indirect utility functions are also identical for the two classes, and therefore we have

$$u_1 = Ar_1(x)^{-b}[w_1(1 - \tau) - qx]^{a+b}, \qquad u_2 = Ar_2(x)^{-b}[1 - \tau - qx]^{a+b}, \quad (26)$$

where $r_i(x)$ is class i's bid rent at x, u_i class i's utility level ($i = 1, 2$), and w_1 class 1's income level. Note that only class 2's income level is normalized as unity, and the proportional income tax imposed on the two classes can no longer be interpreted as a lump-sum tax. Let the total number of residents in class 1 be N_1 and that in class 2 be N_2, and assume that N_1 and N_2 are growing at a common constant rate,

$$\hat{N}_1 = \hat{N}_2 = \mu > 0. \qquad (27)$$

In this case, total tax revenue is

$$S = \tau(w_1 N_1 + N_2), \qquad (28)$$

and the full accommodation condition becomes

$$2\pi g \int_0^{x_1} \frac{x}{h[r_1(x), w_1(1 - \tau) - qx]} dx = N_1,$$

$$2\pi g \int_{x_1}^{x_c} \frac{x}{h[r_2(x), 1 - \tau - qx]} dx = N_2, \qquad (29)$$

where x_1 is the residential boundary between the two classes. Finally, we have the condition that the overall rent function be continuous at x_1 and x_c,

$$r_1(x_1) = r_2(x_1), \qquad r_2(x_c) = 0, \qquad (30)$$

where the opportunity cost of land is still assumed to be zero.

This time we shall define $k \equiv 1/[(w_1 N_1 + N_2)q^2]$. Then it follows from Eqs. (26), (28), and (30) that

$$\hat{k} = 2\lambda(S/J) - \mu = 2\lambda[\tau/\{\pi e(1 - \tau)^2 k\}] - \mu. \qquad (31)$$

This is exactly the same as Eq. (22), and thus the existence, uniqueness, and global stability of a stationary value of k can be proved just as before.

Theorem 9.3   *A balanced growth equilibrium exists and is unique and globally stable in the two-residential class case with different income levels and identical Cobb–Douglas utility functions as well as identical linear transport cost functions, where the opportunity cost of land is zero and transportation invest-ment is financed by a proportional income tax.*

Now let us ask whether our simple golden rule result is still valid. Unlike the previous case of one homogeneous class, it appears difficult to find a simple criterion of optimality in the two-income class case, because we should form a single criterion based on two different utility levels u_1 and u_2. Therefore, the answer appears to depend on the weights to be attached to the utility levels for the two classes. Fortunately, however, we can avoid this complication, because it turns out that both u_1 and u_2 are maximized *simultaneously* when the tax rate is equal to the land share in net income. In fact, using Eq. (26), we can solve Eq. (30) for x_1 and obtain

$$x_1 = (1 - \tau)(w_1 - v)/[q(1 - v)], \qquad \text{where} \quad v \equiv (u_1/u_2)^{1/(a+b)}.$$

Also from Eq. (30) we find $x_c = (1 - \tau)/q$. Then it follows from Eqs. (26) and (29) that

$$
\begin{aligned}
u_1 &= \left[\frac{G}{N_1 q^2} \int_{(1-\tau)(1-w_1)v/(1-v)}^{w_1(1-\tau)} \left[w_1(1 - \tau) - z \right] z^{a/b} \, dz \right]^b \\
&= \left[Hk(1 - \tau)^{2 + a/b} \Phi(v) \right]^b, \\
u_2 &= \left[\frac{K}{N_2 q^2} \int_0^{(1-\tau)(1-w_1)/(1-v)} (1 - \tau - z) z^{a/b} \, dz \right]^b \\
&= \left[Lk(1 - \tau)^{2 + a/b} \Psi(v) \right]^b,
\end{aligned}
\tag{32}
$$

where G, H, K, and L are some positive constants and Φ and Ψ are functions of v. Substituting the equilibrium value of k [given by Eq. (23)] into Eq. (32), we obtain

$$u_1^* = M_1 \tau^b (1 - \tau)^a \phi(v^*), \qquad u_2^* = M_2 \tau^b (1 - \tau)^a \psi(v^*), \tag{33}$$

where M_1 and M_2 are positive constants, ϕ and ψ are functions of v, and $v^* = (u_1^*/u_2^*)^{1/(a+b)}$. Setting $du_1^*/d\tau = du_2^*/d\tau = 0$ (i.e., $dv^*/d\tau = 0$), we find the optimal tax rate $\tilde{\tau}$ to be

$$\tilde{\tau} = b/(a + b). \tag{34}$$

Theorem 9.4 *In the two-income class case with Cobb–Douglas utility functions and zero-opportunity cost of land, the balanced growth equilibrium levels of utility for the two classes are simultaneously maximized when the rate of the proportional income tax is equal to $b/(a + b)$, i.e., the land share in net income.*

This means that with any nonnegative weights attached to u_1 and u_2 to form a single criterion of optimality, our simple golden rule remains valid in the two-income class case. Since the tax rate is the same for the two classes, we may conclude that under the present assumptions there is no conflict between the rich and the poor concerning the fraction of income to be taxed and spent on transportation, provided that a proportional income tax is acceptable to finance transportation investment.

9.4 Modifications

Some modifications are now in order. First, we shall drop the assumption of zero-opportunity cost of land and ask what the optimal tax rate will be if the opportunity cost is positive. Intuition tells us that at each given moment of time, in the positive-opportunity cost case, city size tends to be smaller and therefore a given amount of transportation investment can be utilized more intensively and more efficiently than in the zero-opportunity cost case. This suggests that the optimal tax rate in the case that $r_0 > 0$ is likely to be higher than in the case $r_0 = 0$. In fact, in the case of identical residents the optimal tax rate can be shown to be greater than $b/(a + b)$ when $r_0 > 0$. In that case, we have

$$k^* = \tau/[\lambda^{-1}(\mu/2)\pi e(1 - \tau - y)^2] \tag{35}$$

and

$$u = \left[\frac{F}{Nq^2} \int_y^{1-\tau} (1 - \tau - z)z^{a/b} \, dz\right]^b, \tag{36}$$

where $y = c(r_0)^{b/(a+b)}u^{1/(a+b)}$, and F and c are positive constants. From Eqs. (35) and (36) it follows that

$$u^* = P\tau^b(1 - \tau)^a \left(\frac{1 - \tau}{1 - \tau - y}\right)^{2b} \left[Q - \frac{b}{a + b}\left(\frac{y}{1 - \tau}\right)^{1 + a/b}\right.$$
$$\left. + \frac{b}{a + 2b}\left(\frac{y}{1 - \tau}\right)^{2 + a/b}\right]^b,$$

where P and Q are positive constants. Since $dy/d\tau = 0$ when $du^*/d\tau = 0$, the maximization condition will be

$$\frac{(du^*/d\tau)}{u^*} = \frac{b}{\tau} - \frac{a}{1 - \tau} + 2b\left(\frac{1}{1 - \tau - y} - \frac{1}{1 - \tau}\right)$$
$$+ \frac{by^{1 + a/b}P^{1/b}\tau}{(1 - \tau)(1 - \tau - y)(u^*)^{1/b}} = 0,$$

implying that $b/\tau - a/(1 - \tau) < 0$, i.e., $\tilde{\tau} > b/(a + b)$, when $r_0 > 0$. Thus, the golden rule value $b/(a + b)$ may be regarded as a *lower* limit to the optimal tax rate so that $\tilde{\tau} \to b/(a + b)$ as $r_0 \to 0$.

Theorem 9.5 *In the Cobb–Douglas case with identical residents, the optimal tax rate $\tilde{\tau}$ is greater than $b/(a + b)$, i.e., the land share in net income, when the opportunity cost of land r_0 is positive; and $\tilde{\tau}$ approaches $b/(a + b)$ as r_0 goes to zero.*

Next, in the case of identical residents, we might choose the criterion of optimality

$$\int_0^\infty u \exp(-\delta t)\, dt \tag{37}$$

in the presence of some positive discount rate for future consumption, where δ is a given positive constant. On the assumption that $r_0 = 0$ and

$$g(S/J) = \gamma(S/J)^n, \tag{38}$$

we can explicitly solve for the optimal tax rate as follows. Given the specification (38), the Hamiltonian for the problem of maximizing the objective function (37) subject to

$$\dot{k} = 2\gamma[\tau/\{\pi e(1 - \tau)^2 k\}]^n k - \mu k \tag{39}$$

can be expressed as

$$\mathcal{H} = \{Gk^b(1 - \tau)^{a+2b} + p[2\gamma(\pi e)^{-n}\tau^n(1 - \tau)^{-2n}k^{1-n} - \mu k]\} \exp(-\delta t), \tag{40}$$

where p is a continuous function of time. Then the optimality conditions will be

$$-(a + 2b)Gk^b(1 - \tau)^{a+2b-1} + pn\left(\frac{1}{\tau} + \frac{2}{1 - \tau}\right)2\gamma(\pi e)^{-n}\tau^n(1 - \tau)^{-2n}k^{1-n} = 0, \tag{41}$$

and

$$\dot{p} - (\delta + \mu)p = -[bGk^{b-1}(1 - \tau)^{a+2b} + p(1 - n)2\gamma(\pi e)^{-n}\tau^n(1 - \tau)^{-2n}k^{-n}]. \tag{42}$$

Setting $\dot{k} = \dot{p} = 0$, we find the optimal tax rate $\tilde{\tau}$ as

$$\tilde{\tau} = b/[a + b + \delta(a + 2b)/(\mu n)] \tag{43}$$

from Eqs. (39), (41), and (42). The result (43) might be called the "modified" golden rule of urban transportation investment, and in this case the value $b/(a + b)$ sets an *upper* limit to the optimal tax rate.

Theorem 9.6   *In the Cobb–Douglas case with $r_0 = 0$ and the specification* (38), *the criterion of optimality* (37) *leads to*

$$\tilde{\tau} = b/[a + b + \delta(a + 2b)/(\mu\eta)] < b/(a + b), \tag{44}$$

when the discount rate δ is positive; and

$$\tilde{\tau} \to b/(a + b) \qquad as \quad \delta \to 0. \tag{45}$$

Finally, it should be pointed out that although our analysis in this chapter is focused on long-run balanced growth, the usefulness of our results is not necessarily restricted to long-run analysis. Recalling the well-known "turn-pike theorem" (see, e.g., Samuelson, 1965), we can state that if the criterion of optimality is given by $\int_0^T u \exp(-\delta t)\, dt$, and the planning period T is sufficiently long, then the optimal growth path with given initial and terminal conditions will be arbitrarily close to the "turnpike" and the optimal tax rate will be approximately equal to the modified golden rule value (13) for most of the planning period. In this sense, we may conclude that our golden rule almost always applies in the growing urban economy.

REFERENCES

Miyao, T. (1977). The golden rule of urban transportation investment. *Journal of Urban Economics* **4**, 448–458.
Muth, R. F. (1969). *Cities and Housing.* Univ. of Chicago Press, Chicago.
Phelps, E. S. (1961). The golden rule of accumulation. A fable for growthmen *American Economic Review* **51**, 638–643.
Samuelson, P. A. (1965). A catenary turnpike theorem involving consumption and the golden rule. *American Economic Review* **55**, 486–496.
Solow, R. M. (1973). On Equilibrium models of urban location. In *Essays in Modern Economics* (M. Parkin, ed.), pp. 2 16. Longman Group, London.

CHAPTER **10**

Urban Growth and Unemployment

In this chapter we take account of the phenomenon of urban unemployment in a demand-oriented model of urban growth and examine the effect
of unemployment on the dynamic property of a growing urban economy.[1]
Although in the literature there have been some attempts to construct urban
growth models which incorporate the spatial aspect of urban growth, including ours in the previous two chapters, none of those models has paid
explicit attention to the phenomenon of urban unemployment. On the other
hand, there have been a series of studies on urban unemployment itself,
particularly in the context of economic development (e.g., Harris and Todaro,
1970; Todaro, 1969), but none of them has taken account of the spatial aspect
of urban growth and its implications for the dynamic property of the urban
economy. In short, what is lacking in the existing literature is an analysis of
urban growth with explicit regard to both urban unemployment and spatial
growth, an analysis which can provide a new insight into the workings of a
growing urban economy. We intend to offer such an analysis in this chapter.

First, starting with an equilibrium pattern of individual firms in space, we
derive an aggregate production function for the urban economy as a whole.
Then we construct a demand-oriented model of spatial growth and introduce
dynamic adjustment processes of capital, population, and the wage rate
through time. Finally, dynamic and comparative static analyses are con-
ducted in order to examine the long-run behavior of such variables as unemployment, income, and population in response to exogenous demand
growth for output in the urban economy.

[1] This chapter is based on Miyao (1980).

Our analysis shows that the long-run property of the model depends crucially on the sensitivity of the rate of nominal wage increase to changes in the rate of price increase in the city. If it is relatively insensitive, the unemployment rate will decrease in the long run, when the rate of demand growth for output increases. If it is relatively sensitive, the long-run rate of unemployment tends to rise with the rate of demand growth. In the latter case, however, the city will provide a higher real wage rate which accelerates the rate of population inflow to sustain balanced growth at a faster rate.

Furthermore, the long-run balanced growth equilibrium is shown to be locally stable unless the rate of nominal wage increase is highly sensitive to changes in the rate of price increase. In fact, the stability condition for the present model is weaker than what would be derived from nonspatial aggregate growth models, because of two additional sources of flexibility in the present model, i.e., an adjustment in the supply of output due to spatial expansion or contraction, and an adjustment in the demand for output due to the responsiveness of demand to price changes. Our dynamic analysis also shows that local stability conditions are not strong enough to ensure global stability. However, the balanced growth equilibrium tends to be globally stable in a "full-employment" urban economy, but is less likely to be so in an "underemployment" urban economy.

10.1   The Aggregate Production Function

Let us focus on the spatial aspect of production activity by taking up a von Thünen type model of industrial location. As in Chapter 8, we assume a single group of firms having identical production functions and producing the same kind of commodity in a circular city with a central market to which all products must be transported. Again, unit transport cost is assumed to be a linear function of distance.

Consider a production function depending on three factors, namely, land, capital, and labor. Specifically, a firm which is located at distance x from the center has the Cobb–Douglas production function[2]

$$Y(x) = T(x)^a K(x)^b E(x)^c, \tag{1}$$

[2] Here, the Cobb–Douglas form is assumed for analytical simplicity with no empirical verification, although it could be tested by checking the constancy of factor shares, etc. In the present analysis, the Cobb–Douglas assumption is important, if not necessary, for derivation of an aggregate production function, as will be seen later. It should be added that the assumption of constant returns to scale can be relaxed so as to allow some degree of increasing returns. For details, see footnotes 4–6.

with $a > 0$, $b > 0$, $c > 0$, $a + b + c = 1$, where $Y(x)$, $T(x)$, $K(x)$, and $E(x)$ are output, land, capital, and labor, respectively, at x. Under perfect competition, constant returns to scale lead to zero profit, i.e., product price net of transport cost is equal to unit production cost.

$$1 - qx/p = Cr(x)^a v^b w^c, \tag{2}$$

with $C \equiv a^{-a} b^{-b} c^{-c}$, where q is marginal (unit) transport cost, and p, $r(x)$, v, and w are the price of output, the real rental price of land at x, the real return on capital, and the real wage rate, respectively. Note that p, v, and w do not depend on x, but $r(x)$ does, since it follows from Eq. (2) that

$$r(x) = C^{-1/a} v^{-b/a} w^{-c/a} (1 - qx/p)^{1/a}. \tag{3}$$

Given the Cobb–Douglas production function (1), profit maximization leads to

$$Y(x)(1 - qx/p) = r(x)T(x)/a = vK(x)/b = wE(x)/c.$$

Then, total capital stock K, total labor employment E, and total output Y can be expressed as

$$K = s \int_0^{x_b} x \frac{K(x)}{T(x)} dx = \frac{sb}{av} \int_0^{x_b} xr(x)\, dx, \tag{4}$$

$$E = s \int_0^{x_b} x \frac{E(x)}{T(x)} dx = \frac{sc}{aw} \int_0^{x_b} xr(x)\, dx, \tag{5}$$

$$Y = s \int_0^{x_b} x \frac{Y(x)}{T(x)} dx = \frac{s}{a} \int_0^{x_b} \frac{xr(x)}{1 - qx/p}\, dx, \tag{6}$$

where $s \equiv 2\pi g$, g is a constant fraction of land available for production at each x, and x_b is the distance from the center to the urban boundary.

Let us define total net output Q as total output minus total transport cost,

$$Q = s \int_0^{x_b} x(1 - qx/p)\frac{Y(x)}{T(x)} dx = \frac{s}{a} \int_0^{x_b} xr(x)\, dx. \tag{7}$$

It is easy to see from Eqs. (4), (5), and (7) that

$$Q = vK/b = wE/c, \tag{8}$$

and in view of Eq.(3),

$$Q = (s/a)C^{-1/a} v^{-b/a} w^{-c/a} \int_0^{x_b} x(1 - qx/p)^{1/a}\, dx$$

$$= (1/a)C^{-1/a}(bQ/K)^{-b/a}(cQ/E)^{-c/a} L, \tag{9}$$

where

$$L \equiv s \int_0^{x_b} x(1 - qx/p)^{1/a} \, dx, \tag{10}$$

which might be called the "efficient" stock of land, consisting of all individual pieces of land with the respective weights of efficiency represented by the term[3] $(1 - qx/p)^{1/a}$. From Eqs. (2) and (9), we obtain the aggregate production function for the city as a whole,

$$Q = L^a K^b E^c. \tag{11}$$

Next, we shall see how the urban boundary x_b is determined under perfect competition. In equilibrium, the overall rent function should be continuous at the urban boundary,

$$r(x_b) = 0, \tag{12}$$

where the opportunity cost of land is assumed to be zero for analytical simplicity. Equations (3) and (12) together imply

$$x_b = p/q. \tag{13}$$

Then, defining $z \equiv qx/p$, we can express the efficient stock of land (10) as

$$L = (p/q)^2 s \int_0^1 z(1 - z)^{1/a} \, dz, \tag{14}$$

and find the ratio of total output Y to total net output Q from Eqs. (3), (6), and (9) as

$$\frac{Y}{Q} = \frac{\int_0^{x_b} x(1 - qx/p)^{(1-a)/a} \, dx}{\int_0^{x_b} x(1 - qx/p)^{1/a} \, dx} = \frac{\int_0^1 z(1 - z)^{(1-a)/a} \, dz}{\int_0^1 z(1 - z)^{1/a} \, dz} = B, \tag{15}$$

where B is a positive constant. Thus, from Eqs. (11), (14), and (15), we find the aggregate production function for total output,[4]

$$Y = BL^a K^b E^c = A(p/q)^{2a} K^b E^c, \tag{16}$$

where A is a constant,

$$A = B \left[s \int_0^1 z(1 - z)^{1/a} \, dz \right]^a. \tag{17}$$

---

[3] As pointed out in Chapter 8, this is analogous to the concept of the "efficient" stock of capital in the context of economic growth and technical progress (see, e.g., Solow, 1960).

[4] Note that with $a + b + c = 1$ in Eq. (1) the aggregate production function (16) exhibits constant returns to scale. Increasing returns can be taken into account by assuming $a + b + c > 1$, or by specifying the production function as $Y = Y^e L^a K^b E^c$, i.e., $Y = L^{a/(1-e)} K^{b/(1-e)} E^{c/(1-e)}$, where $0 < e < 1$ and $a + b + c = 1$. Allowing increasing returns to scale would not alter our results in any important ways, however.

10.2   The Model and Balanced Growth

Let us construct a demand-oriented model of urban growth based on the aggregate production function derived in the previous section. Here we are concerned with the type of urban growth caused by the exogenous growth of demand for output in the city. Specifically, we assume that the demand for output is growing at a given constant rate γ, and that the demand is negatively related to the price of output p,

$$D = D_0 p^{-\epsilon} e^{\gamma t}, \qquad D_0 > 0, \quad \epsilon \geqq 0, \quad \gamma \geqq 0. \tag{18}$$

At each moment of time, demand equals supply,

$$D = Y = A(p/q)^{2a} K^b E^c, \tag{19}$$

where unit (marginal) transport cost q is assumed to be increasing or decreasing (or constant) at a given constant rate $\eta \gtreqless 0$; i.e., $q = q_0 e^{\eta t}$.

Following the "neo-Keynesian" literature (e.g., Engle, 1974), we shall introduce some dynamic adjustment processes of capital K, population N, and the nominal wage rate W. First, the rate of capital inflow to the city, defined by $\hat{K} \equiv (dK/dt)/K$, is assumed to be an increasing function of the real rate of return v in the city, given the national (or international) rate of return on capital, say $\bar{v}$,

$$\hat{K} = \kappa(v; \bar{v}), \qquad \kappa_v > 0. \tag{20}$$

Second, the rate of net migration into the city, $\hat{N} \equiv (dN/dt)/N$, depends on the real wage rate w and the unemployment rate, i.e., $u \equiv 1 - (E/N)$, in the city. Naturally, N is increasing in w and decreasing in u, given the national rates of real wage $\bar{w}$ and unemployment $\bar{u}$,

$$\hat{N} = \mu(w, u; \bar{w}, \bar{u}), \qquad \mu_w > 0, \quad \mu_u < 0. \tag{21}$$

It is worth mentioning that a special case of the migration function μ may be written as $\mu[w(1 - u); \bar{w}, \bar{u}]$, which has been widely used in the literature (e.g., Harris and Todaro, 1970; Todaro, 1969). Third, there is assumed to be some imperfection in the labor market within the city, say, due to the presence of local labor unions, which is assumed by Calvo (1978) in the Harris–Todaro model. Following the tradition of the Phillips–curve approach (see Phillips, 1958), we suppose that the rate of increase of the nominal wage rate, $\hat{W} \equiv (dW/dt)/W$, is decreasing in the unemployment rate u and increasing in the rate of price increase, $\theta \equiv (dp/dt)/p$, given the national rates of unemployment $\bar{u}$ and price increase $\bar{\theta}$,

$$\hat{W} = \omega(u, \theta, \bar{u}, \bar{\theta}), \qquad \omega_u < 0, \quad \omega_\theta > 0. \tag{22}$$

Define a balanced growth equilibrium as a state in which all factors L, K, E, and N are growing at a common constant rate. This means that the quantity variables Y, L, K, E, and N will maintain the same proportions to each other over time and, as a result, the real price variables w and v will remain stationary, as will be seen later. In such an equilibrium, we have[5]

$$\gamma - \epsilon\hat{p} = \hat{Y} = 2(\hat{p} - \hat{q}) = \hat{K} = \hat{E} \qquad (23)$$

from Eqs. (18) and (19), and

$$\hat{u} = 0 \qquad (24)$$

from the definition of $u \equiv 1 - (E/N)$ and the condition $\hat{E} = \hat{N}$. It is easy to see from Eqs. (8) and (11) that

$$\hat{v} = 0, \qquad w = 0. \qquad (25)$$

Since $w \equiv W/p$, we note that $\hat{w} = 0$ implies $\hat{W} - \hat{p} = 0$, i.e.,

$$\theta = \hat{W}. \qquad (26)$$

In view of Eqs. (23)–(26), we can determine the balanced growth equilibrium values of θ, v, w, and u by solving the equation system

$$\gamma - \epsilon\theta = 2(\theta - \eta), \qquad (27)$$

$$\gamma - \epsilon\theta = \hat{K} = \kappa(v; \bar{v}), \qquad (28)$$

$$\gamma - \epsilon\theta = \hat{N} = \mu(w, u; \bar{w}, \bar{u}), \qquad (29)$$

$$\theta - \hat{W} = \omega(u, \theta; \bar{u}, \bar{\theta}). \qquad (30)$$

From Eq. (27) the balanced growth equilibrium value of θ is obtained as

$$\theta^* = (\gamma + 2\eta)/(2 + \epsilon). \qquad (31)$$

Given this value of θ, we can find u^* by solving Eq. (30) for u. Then, w^* is found from Eq. (29) with the equilibrium values of θ and u. Also, we can obtain v^* by solving Eq. (28) for v in view of the value of θ^*. In what follows, the existence of such equilibrium values is assumed.

Now we can see how the balanced growth equilibrium values of θ, v, w, and u will change with an increase in the rate of demand growth γ. First, it is clear from Eqs. (28) and (31) that

$$d\theta^*/d\gamma = 1/(2 + \epsilon) > 0, \qquad dv^*/d\gamma = 2/[(2 + \epsilon)\kappa_v] > 0. \qquad (32)$$

---

[5] In the case of increasing returns with $Y = Y^e L^a K^b E^c$, condition (23) should read $\gamma - \epsilon\hat{p} = \hat{Y} = 2(\hat{p} - \hat{q})/(1 - e) = \hat{K}/(1 - e) = \hat{E}/(1 - e)$. In this case, some of the subsequent equations must be modified accordingly. Also see footnote 6.

That is to say, both the rate of price increase and the real return on capital will increase with the rate of demand growth.

The effects of an increase in γ on u^* and w^* are not so straightforward, however. From Eqs. (30) and (32), we find

$$du^*/d\gamma = (1 - \omega_\theta)/[(2 + \epsilon)\omega_u] \gtreqless 0 \qquad \text{as} \quad \omega_\theta \lesseqgtr 1. \qquad (33)$$

In other words, when the rate of demand growth increases, the unemployment rate will increase *or* decrease in the long run, according as a relatively large *or* small change in the rate of nominal wage increase is induced by a marginal change in the rate of price increase in the city. It may be pointed out that the first case, $du^*/d\gamma > 0$, is somewhat similar to the so-called "Todaro paradox" (Todaro, 1969); i.e., the higher the rate of urban industrial expansion, the higher the urban unemployment rate in the long run. It should also be noted that unlike some aggregative models (e.g., Solow and Stiglitz, 1968), we cannot rule out the case $\omega_\theta > 1$, as we shall see later. Next, Eqs. (29), (31), and (33) together yield

$$\frac{dw^*}{d\gamma} = \frac{2}{2 + \epsilon}\left[1 - \frac{\mu_u(1 - \omega_\theta)}{2\omega_u}\right]\frac{1}{\mu_w} \gtreqless 0 \qquad \text{as} \quad \omega_\theta \lesseqgtr 1 - \frac{2\omega_u}{\mu_u}. \qquad (34)$$

From Eqs. (33) and (34) we find the following. (1) Given a sufficiently high value of ω_θ (>1), an increase in the rate of demand growth will lead to a higher rate of unemployment, but at the same time will lead to a higher real wage rate which accelerates the rate of population inflow to sustain balanced growth at a faster rate. (2) Given a sufficiently low value of ω_θ ($<1 - 2\omega_u/\mu_u$), a higher rate of demand growth will reduce both the unemployment rate and the real wage rate. In this case, it is the reduction in the unemployment rate which induces more population inflow. (3) Given a value ω_θ between 1 and $1 - 2\omega_u/\mu_u$, an increase in the rate of demand growth will yield a lower rate of unemployment and a higher real wage rate, both of which contribute to a faster rate of population inflow for sustaining balanced growth.

In summary, we have proved the following.

Theorem 10.1 *In the present Cobb–Douglas case with zero opportunity cost of land, the exponentially growing demand function* (18), *and the dynamic adjustment processes of capital, population, and the wage rate, given by Eqs.* (20)–(22), *we find that in balanced growth equilibrium the rate of price increase* θ^* *and the real return on capital* v^* *will increase with the rate of demand growth* γ,

$$d\theta^*/d\gamma > 0 \qquad and \qquad dv^*/d\gamma > 0, \qquad (35)$$

and the unemployment rate u^* *and the real wage rate* w^* *may increase or de-*

crease with γ, depending on the responsiveness of the rate of wage increase to the rate of price increase in the city,

$$du^*/d\gamma \gtreqless 0 \quad as \quad \omega_\theta \gtreqless 1, \tag{36}$$

$$dw^*/d\gamma \gtreqless 0 \quad as \quad \omega_\theta \gtreqless 1 - (2\omega_u/\mu_u). \tag{37}$$

Finally, it should be added that if the urban economy is sufficiently close to full employment, then $dw^*/d\gamma$ is likely to be positive, since it is natural to assume that $-\omega_u \to \infty$ as $u \to 0$, due to the ordinary shape of the Phillips curve, and therefore it is always true that $\omega_\theta > 0 > 1 - 2\omega_u/\mu_u$. Intuitively speaking, in this case there is not much room left for a reduction in the unemployment rate in order to induce a higher rate of population inflow and thus the real wage rate must increase to do the job instead.

10.3 Local Stability Analysis

We are now in a position to analyze the stability property of the balanced growth equilibrium characterized in the previous section. Here our aim is twofold. First, a true dynamic analysis is provided to help understand the long-run growth and adjustment processes of the urban economy. Second, a stability condition is derived to see whether any of the results obtained so far are associated with dynamic instability. In particular, we shall see if the high responsiveness of the rate of wage increase to the rate of price increase (i.e., $\omega_\theta > 1$) is compatible with dynamic stability or not. If not, the comparative static results obtained in the case $\omega_\theta > 1$ should be considered meaningless on the ground of Samuelson's Correspondence Principle (Samuelson, 1947).

Our dynamic analysis is carried out in terms of three variables

$$z \equiv Y/K, \quad y \equiv Y/E, \quad m \equiv E/N \quad (=1 - u). \tag{38}$$

It follows from Eqs. (8), (11), and (15) that

$$v = bQ/K = (b/B)z, \tag{39}$$

$$w = cQ/E = (c/B)y. \tag{40}$$

Using relations (38)–(40), we can rewrite the dynamic equation system (20)–(22) as

$$\hat{K} = \kappa\big[(b/B)z; \bar{v}\big] \equiv \sigma(z), \qquad\qquad \sigma' > 0, \tag{41}$$

$$\hat{N} = \mu\big[(c/B)y, 1 - m; \bar{w}, \bar{u}\big] = \phi(y, m), \qquad \phi_y > 0, \quad \phi_m > 0, \tag{42}$$

$$\hat{W} = \omega\big[1 - m, \theta; \bar{u}, \bar{\theta}\big] \equiv \psi(m, \theta), \qquad \psi_m > 0, \quad \psi_\theta > 0. \tag{43}$$

In order to derive the fundamental differential equations in terms of z, y, and m, we have from Eqs. (18) and (19) that

$$\gamma - \epsilon\theta = 2a(\theta - \eta) + b\hat{K} + c\hat{E},$$

which can be solved for θ as

$$\theta = (2a\eta + \gamma - b\hat{K} - c\hat{E})/(2a + \epsilon). \tag{44}$$

On the other hand, we find from Eqs. (11) and (40) that

$$w = c(L/E)^a(K/E)^b,$$

and its logarithmic differentiation with respect to time is

$$\hat{W} - \theta = a[2(\theta - \eta) - \hat{E}] + b(\hat{K} - \hat{E}),$$

which, in view of Eqs. (43) and (44) for $\hat{W}$ and θ, respectively, can be solved for $\hat{E}$ as a function of m and $\hat{K}$,

$$\hat{E} = G(m, \hat{K}), \tag{45}$$

such that

$$\partial\hat{E}/\partial m \equiv G_m = (2a + \epsilon)\psi_m/J, \tag{46}$$

$$\partial\hat{E}/\partial\hat{K} \equiv G_k = b(1 - \psi_\theta - \epsilon)/J, \tag{47}$$

where $J = c(\omega_\theta - 1 - 2a) - (a + b)(2a + \epsilon) = c(\psi_\theta - 1) - 2a - (a + b)\epsilon$. Clearly, $J < 0$ if and only if

$$\psi_\theta < 1 + 2a/c + \epsilon(a + b)/c. \tag{48}$$

Since $\psi_m > 0$, it follows from Eq. (46) that $G_m < 0$ if and only if condition (48) holds.[6]

Now the fundamental differential equations can be expressed as

$$\hat{z} = \hat{Y} - \hat{K} = \gamma - \sigma(z), \tag{49}$$

$$\hat{y} = \hat{Y} - \hat{E} = \gamma - G[m, \sigma(z)], \tag{50}$$

$$\hat{m} = \hat{E} - \hat{N} = G[m, \sigma(z)] - \phi(y, m). \tag{51}$$

Here we are concerned with the local stability property of the dynamic system in a small neighborhood of equilibrium. Let z^*, y^*, and m^* denote

---

[6] In the case of increasing returns with $Y = Y^e L^a K^b E^c$, condition (48) becomes

$$\psi_\theta < 1 + 2a/c + \epsilon(a + b)(1 - e)/c.$$

the balanced growth equilibrium values of z, y, and m, respectively. The system (49)–(51) can then be approximated in the linear form

$$
\begin{bmatrix} \dot{z - z^*} \\ y - y^* \\ m - m^* \end{bmatrix} = \begin{bmatrix} -\sigma'z & 0 & 0 \\ -G_k\sigma'y & 0 & -G_m y \\ G_k\sigma'm & -\phi_y m & (G_m - \phi_m)m \end{bmatrix} \begin{bmatrix} z - z^* \\ y - y^* \\ m - m^* \end{bmatrix}, \tag{52}
$$

where the dot denotes differentiation with respect to time and all the variables are evaluated at the equilibrium.

We can show that condition (48) is a necessary and sufficient condition for the linear system (52) to be stable, since condition (48) is necessary and sufficient for every characteristic root of the system (52) to have a negative real part, as will be shown. Because of the block-triangularity of the matrix on the right-hand side of Eq. (52), we find the corresponding characteristic equation to be

$$
(-\sigma'z - \lambda) \begin{vmatrix} -\lambda & -G_m y \\ -\phi_y m & (G_m - \phi_m)m - \lambda \end{vmatrix} = 0, \tag{53}
$$

which in general gives three roots λ satisfying Eq. (53). It is clear that one of the roots is $-\sigma'z < 0$ and that each of the remaining two roots has a negative real part if and only if condition (48) is satisfied, because the determinant in Eq. (53) has the trace $(G_m - \phi_m)m < 0$, and the determinant itself is $-G_m y\phi_y m > 0$ if and only if $G_m < 0$, which is in turn equivalent to condition (48). In view of the definition of ψ in Eq. (43), we have $\omega_\theta \equiv \psi_\theta$, and therefore we have proved the following theorem.

Theorem 10.2 *Under the present assumptions with the dynamic processes* (20)–(22), *the balanced growth equilibrium is locally stable if and only if*

$$
\omega_\theta < 1 + 2a/c + \epsilon(a + b)/c. \tag{54}
$$

This theorem implies that in Theorem 10.1, it is meaningful to consider the case $\omega_\theta > 1$ since it is compatible with stability so long as condition (54) is met. This result is in contrast to the result obtained from some aggregative models (e.g., Solow and Stiglitz, 1968) in which $\omega_\theta > 1$ does lead to instability.

An economic interpretation of our stability result may be given as follows. Suppose that, starting at the balanced growth equilibrium, the growth rate of demand for output happens to exceed that of supply, and as a result the rate of price increase is greater than its equilibrium value. In the case $\omega_\theta > 1$, the nominal wage rate will respond to the price increase more than proportion-

TABLE 10.1

Case		$du^*/d\gamma$	$dw^*/d\gamma$
1	$1 < \omega_\theta < 1 + 2a/c + \epsilon(a + b)/c$	+	+
2	$\omega_\theta = 1$	0	+
3	$1 - 2\omega_u/\mu_u < \omega_\theta < 1$	−	+
4	$\omega_\theta = 1 - 2\omega_u/\mu_u$	−	0
5	$0 < \omega_\theta < 1 - 2\omega_u/\mu_u$	−	−

ately, so that the real wage rate tends to rise and labor employment tends to fall. The reduction in employment would decelerate the growth of product supply, which would in turn widen the initial gap between supply and demand and would result in instability, *if* nothing happens to offset the reduction in employment. In our present model, something does happen; in fact, there are two forces which will reduce the initially assumed excess demand for output. First, notice that the higher the rate of price increase, the more land will be profitably utilized, i.e., the higher the rate of growth of the efficient stock of land, as seen in Eq. (14). This increase in land tends to offset the decrease in labor and to reduce the excess demand for output by increasing the growth rate of supply. Actually, this is the case if the percentage increase in output due to the price increase $2a\hat{p}$ is greater than the percentage decrease in output due to the reduction in labor employment cw, i.e., if $\hat{w}/\hat{p} < 2a/c$, which is reflected in the second term of the stability condition (54). Second, an additional source of flexibility is the price elasticity of demand for output. As the rate of price increase rises by one percent, say, the growth rate of demand will fall by ϵ percent, due to Eq. (18), and as a result the excess demand for output tends to be reduced. This is reflected in the third term of condition (54). Note that the more price elastic the demand for output (the greater the value of ϵ), the less strict is the stability condition, as is naturally expected. Our final result regarding the responses of the unemployment rate and the real wage rate to the rate of demand growth is summarized in Table 10.1.

10.4 Global Stability

Now we examine the global stability property of the balanced growth equilibrium in order to see how the urban economy will adjust its growth path for a large initial displacement. Here our argument is brief and rather intuitive.

By reviewing the dynamic system (49)–(51), we notice that Eq. (49) is a single differential equation involving only one variable z and that, in view of the fact that $\sigma' > 0$, the balanced growth equilibrium value of $z = \sigma^{-1}(\gamma)$ will be asymptotically reached for any given initial value of z, regardless of what happens to the other variables y and m. Thus, for the purpose of finding the long-run dynamic motion of y and m, we may proceed with two differential equations (50) and (51) including only two variables y and m by fixing the value of z at its equilibrium value $z^* = \sigma^{-1}(\gamma)$,

$$\hat{y} = \gamma - G(m, \gamma), \tag{55}$$

$$\dot{m} = G(m, \gamma) - \phi(y, m). \tag{56}$$

The global stability property of the system can best be analyzed by using a phase diagram as in Fig. 10.1, where m^* is obtained by solving the equation $0 = y = \gamma - G(m, \gamma)$ for m, and the slope of the curve representing $\dot{m} = 0$ is

$$(dy/dn)_{m=0} = (G_m - \phi_m)/\phi_y < 0. \tag{57}$$

As seen in Fig. 10.1, the global stability of the balanced growth equilibrium is not ensured in general, since we cannot exclude the possibility of divergent spiral or limit cycle movements for a large initial displacement.

Theorem 10.3 *Under the present assumptions including the local stability condition (54), the balanced growth equilibrium may not be globally stable, and per capita income y and the employment rate m (thus, w and u as well) tend to exhibit cyclical fluctuations around their respective equilibrium values,*

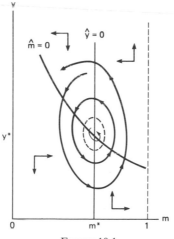

FIGURE 10.1

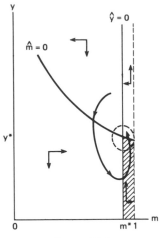

FIGURE 10.2

while the output–capital ratio z and the rate of return on capital v will approach their equilibrium values monotonically after a large initial disturbance.

Finally, it should be pointed out that if in the long run the urban economy is capable of attaining an employment rate sufficiently close to full employment, the adjustment process of the economy is likely to be globally stable. This is because with m^* sufficiently close to unity, the shaded area in Fig. 10.2 becomes so thin that any path must go into a stable region in the neighborhood of equilibrium. If, on the other hand, the economy has a long-run employment rate which is far below unity, there is no such presumption of global stability. This argument may suggest an interesting contrast between the global stability of a long-run–*full-employment* urban economy and the global instability of a long-run–*underemployment* urban economy.

REFERENCES

Calvo, G. A. (1978). Urban unemployment and wage determination in LDC's: Trade unions in the Harris–Todaro model. *International Economic Review* **19,** 65–81.
Engle, R. F. (1974). Issues in the specification of an econometric model of metropolitan growth. *Journal of Urban Economics* **1,** 250–267.
Harris, J. R., and Todaro, M. (1970). Migration, unemployment and development: A two-sector analysis. *American Economic Review* **60,** 126–142.
Miyao, T. (1980). Dynamics of metropolitan growth and unemployment. *Journal of Urban Economics* **8,** 222–235.

Phillips, A. W. (1958). The relation between unemployment and the rate of change of money wage rates in the United Kingdom, 1861–1957. *Economica* **25,** 283–299.

Samuelson, P. A. (1947). *Foundations of Economic Analysis.* Harvard Univ. Press, Cambridge, Massachusetts.

Solow, R. M. (1960). Investment and technical progress. In *Mathematical Methods in the Social Sciences* (K. J. Arrow, S. Karlin, and P. Suppes, eds.), pp. 89–104. Stanford Univ. Press, Stanford, California.

Solow, R. M., and Stiglitz, J. E. (1968). Output, employment and wages in the short run. *Quarterly Journal of Economics* **82,** 537–560.

Todaro, M. (1969). A model of labor migration and urban unemployment in less developed countries. *American Economic Review* **59,** 138–148.

CHAPTER 11

Dynamics of Rural–Urban Migration

In this chapter a kind of externality is introduced, namely, agglomeration economies in production, and the effect of such economies on the dynamic process of rural–urban migration is investigated.[1] We consider a two-sector model of the Harris–Todaro type with manufacturing production in the urban sector and agricultural production in the rural sector, where agglomeration economies are present in the urban manufacturing sector (see Harris and Todaro, 1970: Todaro, 1969). In addition to the emphasis on agglomeration economies, our model takes explicit account of migrants' expectations and migration decisions on the basis of the quantitative choice theory explained in Chapter 6.

We define a long-run steady-state equilibrium as a state of constant urban population and investigate its existence, uniqueness, and stability. It is shown that with agglomeration economies there are likely to exist multiple equilibria, some of which are unstable in the long run. Then our model is able to explain the dynamic process of rural–urban migration, starting from a "rural-economy" equilibrium and moving toward an "urban-economy" equilibrium. We point out that higher expectations of migrants about urban wages and employment, beyond a certain limit, will give rise to a sudden change in the dynamic property of the economy and cause it to move from a rural equilibrium to an urban equilibrium. In fact, it has been shown in Part 2 that the equilibrium may not be unique or stable in the presence of

---

[1] This chapter is based on Miyao and Shapiro (1979).

certain types of externalities. Our result in this chapter may thus be regarded as an example of such effects of externalities on urban growth processes.

In the literature on internal migration, Harris and Todaro (1970), Todaro (1969), and others have focused on rural–urban migration and unemployment in a steady-state equilibrium, but failed to explain how the process of continual migration takes off from a condition of rural stagnation to an urbanized steady state. On the other hand, a large body of literature on economic development has been addressed to the problem of the takeoff process of a developing economy (e.g., Fei and Ranis, 1961), but with no explicit attention paid to the dynamic migration process on the basis of individual optimization behavior, as in the Harris–Todaro model. Our work in this chapter can be viewed as an attempt to fill the gap between the theory of internal migration of the Harris–Todaro type and the theory of economic development based on the takeoff process.

11.1   The Model with Agglomeration Economies

First, as in the typical Harris–Todaro model, we assume that in the economy there are two sectors, the rural and the urban, and the rural sector has a production function with diminishing labor productivity. Specifically,

$$Q_r = (N_r)^a, \qquad 0 < a < 1, \tag{1}$$

where Q_r and N_r are total output and total employment (population) in the rural sector. In this sector there is no unemployment by assumption.

Let us depart from the Harris–Todaro model by introducing agglomeration economies in production in the urban sector:

$$Q_u = (Q_u)^b (E_u)^c, \qquad 0 < b < 1, \quad 0 < c < 1, \quad b + c > 1, \tag{2}$$

where Q_u and E_u are total output and total employment in the urban sector. Since each competitive firm is assumed to be too small to influence the degree of agglomeration economies $(Q_u)^b$ by itself, the urban wage rate in terms of the urban product is equal to the marginal product of labor $\partial Q_u / \partial E_u$ given $(Q_u)^b$, i.e.,

$$w_u = c(Q_u)^b (E_u)^{c-1} = c Q_u / E_u. \tag{3}$$

It follows from Eq. (2) that

$$Q_u = (E_u)^g, \qquad g \equiv c/(1 - b) > 1, \tag{4}$$

and thus from Eq. (3) that

$$w_u = c(E_u)^h, \qquad h \equiv g - 1 > 0. \tag{5}$$

On the other hand, the rural wage rate in terms of the urban product is determined from Eq. (1) as

$$w_r = p\, \partial Q_r/\partial N_r = pa(N_r)^{a-1} = paQ_r/N_r, \tag{6}$$

where w_r is the rural wage rate in terms of the urban product and p is the relative price of the rural product to the urban product. Assuming the aggregate utility function to be Cobb–Douglas,

$$U = (Q_r)^\alpha (Q_u)^{1-\alpha}, \qquad 0 < \alpha < 1, \tag{7}$$

we find

$$pQ_r/Q_u = s \equiv \alpha/(1 - \alpha), \tag{8}$$

and thus Eqs. (1), (4), and (6) lead to

$$w_r = a(N_r)^{a-1} sQ_u/Q_r = as(E_u)^g/N_r. \tag{9}$$

Define the employment rate in the urban sector as

$$m \equiv E_u/N_u, \tag{10}$$

where N_u is total labor force (population) in the urban area. In order to focus on the dynamic process of migration, we simply assume that m is given as

$$0 < m < 1. \tag{11}$$

Furthermore, the total population N for the economy as a whole is assumed to be given and constant so that

$$N = N_r + N_u. \tag{12}$$

Then we can express the urban and rural wage rates w_u and w_r as functions of N_u alone,

$$w_u = cm^h(N_u)^h, \qquad w_r = asm^g(N_u)^g/(N - N_u). \tag{13}$$

Now let us assume that each individual, currently working in the rural sector, has an opinion about the expected urban wage rate z (the employment rate times the urban wage rate) which he would receive if he migrates to the city. We further assume that individuals have different opinions about their expected urban wage rates in such a way that the percentage of rural population with the expected urban wage rate z can be represented by a probability density function

$$f(z) \geq 0, \tag{14}$$

where

$$\int_{-\infty}^{\infty} f(z)\, dz = 1. \tag{15}$$

In deciding whether to migrate to the city, each individual compares what he thinks the expected urban wage rate would be with the rural wage which he is currently earning.[2] An individual will move to the city, if his expected urban wage rate z exceeds his rural wage w_r plus moving cost q, where q is assumed to be given. On the other hand, he will stay in the rural area, if $z < w_r + q$. Then, the proportion of rural population migrating to the urban area is given by

$$P = \int_{w_r + q}^{\infty} f(z)\, dz. \tag{16}$$

It seems reasonable to assume that the mean μ of the distribution of individuals' expected urban wage rates is an increasing function of the actual value of the expected urban wage rate, i.e., the urban wage rate times the urban employment rate. Specifically,

$$\mu = \lambda w_u m, \qquad \lambda > 0. \tag{17}$$

If λ is large (small), particularly larger (smaller) than unity, "general expectations" about urban wages and employment are said to be high (low).

Now consider the counterflow of migrants from the urban area to the rural area. For analytical simplicity, we shall suppose that those who are currently employed in the urban sector will stay there and will not return to the rural area, regardless of the wage rates or the urban employment rate. On the other hand, those who are currently unemployed will compare the rural wage rate with the expected urban wage which they think they could earn in the immediate future if they stay in the urban area. Thus, an individual will migrate to the rural area if his z falls short of the rural wage rate w_r minus moving cost q (then, he would be better off in the rural area even after paying the moving cost), but he will stay in the urban area if $z > w_r - q$. Using the same notation $f(z)$ for the number (density) of unemployed in the urban area with the expected urban wage rate z, we find the proportion of the unemployed who migrate to the rural area to be

$$R = \int_{-\infty}^{w_r - q} f(z)\, dz. \tag{18}$$

Again, it is assumed that the mean ω of the distribution of the unemployed individuals' expected urban wage rates z is proportional to the actual value of the expected urban wage rate $w_u m$:

$$\omega = \theta w_u m, \qquad \theta > 0. \tag{19}$$

In general, θ is not equal to λ in Eq. (17).

The net flow of migration from the rural area to the urban area can then

---

[2] Here we are applying the qualitative choice theory explained in Chapter 6.

be found by subtracting the number of unemployed in the city who migrate to the rural area from the number of rural population migrating to the urban area. Since the former is equal to $R(1 - m)N_u$ and the latter is PN_r, the net migration flow to the city will be

$$\Delta N_u = PN_r - R(1 - m)N_u = P(N - N_u) - R(1 - m)N_u, \qquad (20)$$

in view of Eq. (12). Defining $\dot{N}_u \equiv dN_u/dt$, we have a continuous version of the dynamic Eq. (20) as

$$\dot{N}_u = P(N - N_u) - R(1 - m)N_u. \qquad (21)$$

Since P and R depend only on w_r and w_u, which in turn are functions of N_u alone from Eq. (13), we have a single differential equation with a single variable N_u to determine the dynamic path of urban population N_u with any given initial condition.

11.2 Properties of Equilibrium

Let us define a long-run steady-state equilibrium as a state in which the net migration from the rural area to the urban area is zero. Setting $\dot{N}_u = 0$, or

$$0 = G(N_u) \equiv P(N - N_u) - R(1 - m)N_u, \qquad (22)$$

we may solve for the steady-state equilibrium value of N_u.

The existence of an equilibrium value of N_u can be readily shown, since G, as defined in Eq. (22), is a continuous function of N_u, and

$$G(0) = P(0)N \geqq 0, \qquad G(N) = -R(N)(1 - m)N < 0, \qquad (23)$$

with

$$P(0) \equiv \lim_{N_u \to 0} P = \int_q^{\infty} f(z)\, dz \geqq 0$$

$$\text{with mean} \quad \mu(0) = \lim_{N_u \to 0} \lambda w_u m = 0, \qquad (24)$$

$$R(N) = \lim_{N_u \to N} R = \int_{-\infty}^{\infty} f(z)\, dz = 1$$

$$\text{with mean} \quad \omega(N) = \lim_{N_u \to N} \theta w_u m = \theta \bar{w} m, \qquad (25)$$

where $\bar{w} \equiv cm^h N^h$. It is obvious from Eq. (24) that we shall find an interior solution $0 < N_u^* < N$ if we assume

$$f(z) > 0 \qquad \text{for all} \quad z, \qquad (26)$$

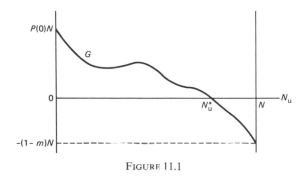

FIGURE 11.1

because in this case

$$G(0) = P(0)N > 0, \qquad G(N) = -R(N)(1 - m)N < 0, \qquad (27)$$

as illustrated in Fig. 11.1.

Theorem 11.1 *Under the present assumptions (1)–(19), a steady-state equilibrium exists; i.e., condition (22) yields an equilibrium value of N_u such that $0 \leq N_u^* < 1$, and furthermore $0 < N_u^* < 1$ if we assume*

$$f(z) > 0 \qquad \text{for all} \quad z. \qquad (28)$$

Next, we shall see if equilibrium is unique, as implicitly assumed in the Harris–Todaro model. It turns out that we may well have multiple equilibria in our model with agglomeration economies in the urban sector. To show the possibility of multiple equilibria in a simple way, let us assume that the mean of the distribution of expected urban wage rates among rural population is always the same as that of the unemployed in the urban sector,[3] i.e.,

$$\lambda = \theta, \qquad \text{or equivalently} \qquad \mu = \omega, \qquad (29)$$

and also suppose that the density function $f(z)$ depends on $(z - \mu)$, so that we can write

$$f(z) = f(z - \mu), \qquad (30)$$

as in the normal distribution case. Then, it follows from Eqs. (16) and (18) that

$$P_r \equiv \partial P/\partial w_r = -f(w_r + q - \mu), \qquad P_\mu \equiv \partial P/\partial \mu = f(w_r + q - \mu),$$
$$R_r \equiv \partial R/\partial w_r = f(w_r - q - \mu), \qquad R_\mu \equiv \partial R/\partial \mu = -f(w_r - q - \mu). \qquad (31)$$

---

[3] This is purely for the sake of analytical simplicity. Essentially the same analysis can be carried out without this assumption.

We are now ready to examine the uniqueness property of equilibrium. Differentiation of G with respect to N_u gives

$$G'(N_u) = [P_r(dw_r/dN_u) + P_\mu(d\mu/dN_u)](N - N_u) - P$$
$$- [R_r(dw_r/dN_u) + R_\mu(d\mu/dN_u)](1 - m)N_u - R(1 - m)$$
$$= m^g(N_u)^{h-1}[\lambda hc - as\{g[N_u/(N - N_u)] + [N_u/(N - N_u)]^2\}]$$
$$\times [f(w_r + q - \mu)(N - N_u) + f(w_r - q - \mu)(1 - m)N_u]$$
$$- P - R(1 - m). \qquad (32)$$

The first term after the second equality above can be shown to be positive *or* negative, according as $0 < N_u < \bar{N}_u$ or $\bar{N}_u < N_u < N$, where $\bar{N}_u$ is given by defining $y \equiv N_u/(N - N_u)$ and solving

$$\lambda hc - as(gy + y^2) = 0 \qquad (33)$$

for a positive root

$$y^* = [-g + \sqrt{g^2 + 4\lambda hc/(as)}]/2 > 0, \qquad (34)$$

and

$$\bar{N}_u \equiv [y^*/(1 + y^*)]N. \qquad (35)$$

Thus

$$\lambda hc - as\{g[N_u/(N - N_u)] + [N_u/(N - N_u)]^2\} \gtrless 0 \qquad \text{as} \quad N_u \gtrless \bar{N}_u. \quad (36)$$

Then it follows from Eq. (32) that

$$G'(0) < 0,$$
$$G'(N_u) \lesseqqgtr 0 \qquad \text{for} \quad 0 < N_u < \bar{N}_u, \qquad (37)$$
$$G'(N_u) < 0 \qquad \text{for} \quad N_u \gtreqqless \bar{N}_u.$$

It is possible to have $G'(N_u) > 0$ for some range of N_u between 0 and $\bar{N}_u$, and to have multiple equilibria, as illustrated in Fig. 11.2. The equilibrium e_1 with the smallest value of N_u might be regarded as the "rural" stagnation equilibrium, whereas the equilibrium e_3 with the largest value of N_u might be regarded as the "urbanized" steady-state equilibrium. The dynamic stability nature of the system can be easily seen from our fundamental dynamic equation (21). In Fig. 11.2, two equilibria e_1 and e_3 are dynamically stable, whereas one equilibrium e_2 is dynamically unstable. It is obvious from Fig. 11.1 that if equilibrium is unique, it is dynamically stable.

In summary, we have shown the following proposition.

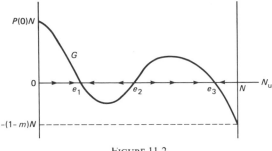

FIGURE 11.2

Theorem 11.2 *Under the present conditions including assumptions (29) and (30), there may well be multiple equilibria, and* (i) *if there are multiple equilibria, some of them are dynamically unstable, and* (ii) *if there is a unique equilibrium, it is dynamically stable.*

11.3 Dynamics of Migration

We are now in a position to explain how the economy will take off, starting from the rural stagnation equilibrium and moving toward the urban steady-state equilibrium. In our model, the most important cause for such a takeoff process seems to be higher expectations about urban wages and urban employment, mainly on the part of rural population and possibly on the part of the urban unemployed. Here, higher expectations mean a higher value of λ.

To see this, we differentiate G partially with respect to λ, and find

$$\partial G/\partial\lambda = [f(w_r + q - \mu)(N - N_u) + f(w_r - q - \mu)(1 - m)N_u]w_u m > 0$$

$$\text{for} \quad 0 < N_u < N. \quad (38)$$

That is to say, a higher value of λ will shift the G curve upward everywhere between 0 and N in Fig. 11.3. Notice that as the G curve shifts up with a higher value of λ, the stable "rural" equilibrium e_1 becomes closer to the unstable equilibrium e_2, and eventually e_1 and e_2 will coincide to become unstable on the right side. That is when the economy will suddenly take off from the rural equilibrium and move toward the urban equilibrium. It should be stressed that such a sudden change in the dynamic property of the system can be caused solely by changes in individuals' subjective expectations with other parameters such as the employment rate unchanged. We can readily generalize this result and obtain the following theorem.

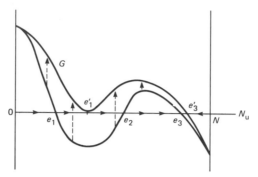

FIGURE 11.3

Theorem 11.3 *Given multiple equilibria under the present conditions includ-ing assumptions (29) and (30), higher expectations about urban wages and urban employment (a higher value of λ) on the part of rural population as well as the urban unemployed will suddenly make the rural equilibrium with the smallest value of N_u dynamically unstable and cause the economy to move toward the stable urban equilibrium with the largest value of N_u.*

The importance of expectational changes may be underscored by pointing out that without a change in the expectational parameter λ, a higher urban employment rate may shift the G curve upward or downward, depending on the values of various parameters, since

$$\partial G/\partial m = -\{gasm^{g-1}(N_u)^g(N-N_u)^{-1} - \lambda[w_u + hcm^h(N_u)^h]\}$$
$$\times f(w_r + q - \mu)(N - N_u) + f(w_r - q - \mu)(1 - m)N_u\} + RN_u,$$

where the first term is positive *or* negative, according as

$$N_u/(N - N_u) \gtrless y^{**} \equiv \lambda(1 + h)c/(gas),$$

and y^{**} has no definite relation to y^* in Eq. (33). Therefore, it is possible to have $\partial G/\partial m < 0$: a downward rather than upward shift in the G curve.[4]

One policy implication which can be drawn from our analysis is that in order to avoid "excessive" rural–urban migration in the presence of urban unemployment, as suggested by Todaro (1969), the government might design policies to lower expectations about urban wages and urban employment on the part of rural population, e.g., by giving them rather "pessimistic" infor-

---

[4] This result seems consistent with the so-called "Todaro paradox" that a higher rate of urban unemployment may result in an accelerated rate of rural–urban migration, because in the present case with $\partial G/\partial m < 0$ a higher rate of urban unemployment will shift the G curve upward and could trigger the takeoff process of rural–urban migration.

mation about urban life publicly and persistently. Such policies may have little effect on migration in the beginning, but as far as such policies affect people's expectations, beyond a certain point, the dynamic flow of migration can be reversed *suddenly*, just as it can be started suddenly due to higher expectations.

REFERENCES

Fei, J. C. H., and Ranis, G. (1961). A theory of economic development. *American Economic Review* **51**, 533–565.

Harris, J. R., and Todaro, M. (1970). Migration, unemployment and development: A two-sector analysis. *American Economic Review* **60**, 126–142.

Miyao, T., and Shapiro, P. (1979). Dynamics of rural–urban migration in a developing economy. *Environment and Planning A* **11**, 1157–1163.

Todaro, M. (1969). A model of labor migration and urban unemployment in less developed countries. *American Economic Review* **59**, 138–148.

PART **4**

CONGESTION AND AGGLOMERATION

CHAPTER **12**

Land Use in a Square City

So far we have assumed that (1) the city under investigation is circular with all roads laid out radially, (2) everyone must travel to the city center, and (3) transport cost is a function of distance alone. Although these assumptions have greatly simplified our analysis, they are rather restrictive and unrealistic in many cases, particularly when we observe large cities in the western United States, where cities are often square in shape, business activities are somewhat dispersed over all of the city area, and transportation costs depend on the level of traffic congestion as well as pure distance to travel. To be more realistic in this direction, in this chapter we present a model of a square city with a grid road system subject to traffic congestion, where business activity is uniformly distributed within the city. The model proves to be useful to investigate optimal land use and optimal city size in relation to congestion costs and agglomeration economies.[1]

In the literature on optimal land use, Solow and Vickrey (1971) have analyzed the optimal allocation of urban land to transportation in a one-dimensional city case and have shown that land-use decisions based on the market value of land would lead to excessive road-building, especially near the city center. A purpose of the analysis in this chapter is to reexamine their conclusion in a square city case, which may be regarded as a natural extension of the Solow–Vickrey city into two-dimensional situations.[2] This is

[1] This chapter is an extension of Miyao (1978).

[2] The Solow–Vickrey analysis has been extended in the context of circular cities by several authors (e.g., Kanemoto, 1975, 1977; Kraus, 1974; Robson, 1976).

151

not an easy task, because in general it is impossible to characterize "equilibrium" traffic patterns which are consistent with individual route choice behavior in two-dimensional cities with alternative routes available to individual road users. Fortunately, however, we can find a simple equilibrium traffic pattern in a special class of cases in which road width is constant everywhere in the city. Focusing on this special case, we shall maintain that land-use decisions using competitive rents would tend to create too many roads in our square city, just as in the Solow–Vickrey linear city; but this tendency is stronger near the city boundaries than at the city center, contrary to the conclusion drawn from the linear city.

Then, we introduce agglomeration economies which are related to city size, and investigate the problem of excessive city size under competition in comparison with optimal city size. It is proved that optimal city size is uniquely determined and that free entry of competitive firms yields an excessively large city, if the degree of agglomeration economies, measured by the elasticity of land productivity with respect to city size, is relatively small.

12.1 The Model and Equilibrium Traffic Patterns

Imagine a two-dimensional rectangular city, including a square-shaped city as a special case, where roads are laid out densely and orthogonally in east–west and north–south directions; they are called "horizontal" and "vertical" roads, respectively. The city is of size $M \times N$, where M is the "horizontal" length and N is the "vertical" length of the city. Place the origin $(0, 0)$ at the city center and consider a point $(x. y)$ at horizontal coordinate x and vertical coordinate y within the city. In the "unit" area $dx\,dy$ at (x, y), the width of the horizontal road is $[1 - m(x, y)]\,dy$ and that of the vertical road is $[1 - n(x, y)]\,dx$, where $m(x, y)\,dy$ and $n(x, y)\,dx$ are the vertical length and the horizontal length, respectively, of the business area at (x, y). Following Solow and Vickrey (1971), we assume that a given volume of traffic, say g, is moved from one unit of business area to any other unit of business area in the city. This means that each unit of business area generates the volume of traffic $gm(x, y)n(x, y)MN$, which has destinations uniformly distributed over all other units of business area.

Let $v_x(x, y)$ (resp. $v_y(x, y)$) be the volume of traffic on the horizontal (resp. vertical) road at (x, y), and $\delta_x(x, y)$ (resp. $\delta_y(x, y)$) be the density of traffic on the horizontal (resp. vertical) road at (x, y). By definition, we have

$$\delta_x(x, y) = v_x(x, y)/[1 - m(x, y)],$$
$$\delta_y(x, y) = v_y(x, y)/[1 - n(x, y)]$$

(1)

for all (x, y) within the city. Furthermore, let $q_x(x, y)$ (resp. $q_y(x, y)$) be the transport cost incurred by individual road users per unit of traffic volume on the horizontal (resp. vertical) road at (x, y). We shall assume that $q_x(x, y)$ (resp. $q_y(x, y)$) depends solely upon the density of traffic on the horizontal (resp. vertical) road at (x, y), i.e.,

$$q_x(x, y) = q[\delta_x(x, y)], \qquad q_y(x, y) = q[\delta_y(x, y)], \qquad (2)$$

where $q(\ ) \geqq 0$, $q'(\ ) > 0$. Note that the functional form of q is assumed to be the same for both horizontal and vertical traffic. This assumption may be justified as an approximation to a city where traffic is regulated by a signal system with "green time" equally allocated to horizontal and vertical traffic at cross roads.

The first task here is to find equilibrium traffic patterns which are consistent with individual route choice behavior in the sense that every road user chooses a cost-minimizing route between the points of his traffic origin and destination so that the present traffic pattern in the city can be self-sustained. Although it is analytically impossible to characterize such equilibrium patterns for all conceivable road configurations in the two-dimensional rectangular city, at least one special class of cases will permit us to give a simple equilibrium traffic pattern explicitly. Specifically, we prove the following propositition.

Theorem 12.1 *In the case that road width is constant and identical horizontally as well as vertically, i.e., $m(x, y) = m$ and $n(x, y) = n$ for all (x, y), there is an equilibrium traffic pattern such that between any pair of traffic origin and destination the two routes along the boundaries of the smallest rectangle that includes the points of origin and destination at its corners are exclusively and equally used by individual road users.*

Proof Consider a point, say E, at (x, y) in Fig. 12.1. Since each unit of business area is both the origin and destination of traffic, the total volume of

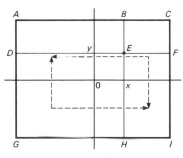

FIGURE 12.1

traffic to be moved between the area represented by the line *DE* and the rectangle *BCIH* and between the line *EF* and the rectangle *ABHG* is equal to

$$2g[mn(\tfrac{1}{2}M + x)mnN(\tfrac{1}{2}M - x) + mn(\tfrac{1}{2}M - x)mnN(\tfrac{1}{2}M + x)].$$

Because of the assumption of equal utilization of two routes between any points, one half of this volume will pass through *E* horizontally. We can then write

$$v_x(x, y) = v_x(x) = 2g(mn)^2 N[(\tfrac{1}{2}M)^2 - x^2], \tag{3}$$

where $v_x(x)$ means that v_x depends only on x and not on y, given $g, m, n, M,$ and N. Similarly, for vertical traffic, we find

$$v_y(x, y) = v_y(y) = 2g(mn)^2 M[(\tfrac{1}{2}N)^2 - y^2]. \tag{4}$$

Thus, from (1) and (2), δ_x and q_x (resp. δ_y and q_y) depend only on x (resp. y),

$$q_x(x) = q[\delta_x(x)] = q[v_x(x)/(1 - m)],$$
$$q_y(y) = q[\delta_y(y)] = q[v_y(y)/(1 - n)]. \tag{5}$$

Finally, it is not hard to see that with the cost pattern given by (5), every road user is actually choosing a route which minimizes total transport cost between the points of his traffic origin and destination. To show this, consider an individual moving from (x_1, y_1) to (x_2, y_2), say, and suppose $x_1 \leqq x_2$ only for the sake of illustration. According to the present route choice pattern, his total transport cost of moving from x_1 to x_2 horizontally is $\int_{x_1}^{x_2} q_x(x)\, dx$, which is *equal* to the total cost of moving from x_1 to x_2 horizontally, not necessarily at once, in any other route involving no horizontal move away from x_2, and is *less* than that in any other route involving some horizontal move away from x_2. This is so because $q_x(x)$ is independent of y, i.e., the cost of *identical* horizontal movement is the same regardless of the vertical coordinate at which the horizontal movement takes place. A symmetric argument applies to the vertical movement from y_1 to y_2, and the theorem is proved.   Q.E.D.

12.2   The Competitive Rent Profile

Let us analyze the land rent profile to be determined by the competitive market in the special class of cases just considered. For the sake of comparison, we shall make the same assumptions as those of Solow and Vickrey (1971): (1) there are no congestion tolls levied on road use, (2) all transport costs are absorbed at points of traffic origin, and (3) the opportunity cost of land is zero. Under these assumptions, the competitive rent profile will reflect the locational advantages of land due to differential transport costs.

It is easy to see that given the foregoing equilibrium traffic pattern, the total transport cost $T(x, y)$ absorbed by a unit of business area at (x, y) may be separated into two parts as

$$T(x, y) = T_x(x) + T_y(y),$$

where $T_x(x)$ and $T_y(y)$ are the portions of total transport cost due to horizontal and vertical movements, respectively, from (x, y) to all other points in the city. In fact, we can calculate transport cost differentials $T'_x(x)$ and $T'_y(y)$ as follows. To find $T'_x(x)$, first consider two points E at (x, y) and E' at $(x + dx, y)$ as in Fig. 12.2. Starting at E' rather than E, one has to bear some extra cost of crossing a thin strip (BH) westward between x and $x + dx$ once for each trip from E' to the rectangle $ABHG$. Then the total of such extra cost is

$$g(mn)^2 N[(\tfrac{1}{2}M) + x]q_x(x), \tag{6}$$

in view of the fact that the total volume of traffic from E' to $ABHG$ is $g(mn)^2 N[(\tfrac{1}{2}M) + x]$ and the unit cost of crossing the thin strip at x is $q_x(x)$ anywhere on the strip BH. On the other hand, starting at E' rather than E, one can save the cost of crossing the strip BH eastward between x and $x + dx$ once for each trip from E' to the rectangle $BCIH$. The total cost to be saved this way is $g(mn)^2 N[(\tfrac{1}{2}M) - x]q_x(x)$. By subtracting this amount from expression (6), we obtain

$$T'_x(x) = 2g(mn)^2 Nxq_x(x), \tag{7}$$

and similarly we can derive

$$T'_y(y) = 2g(mn)^2 Myq_y(y).$$

Since no opportunity cost of land implies no rents for the least-favored locations which are the corner points of the city, the competitive rent $R(x, y)$ at (x, y) can be written as

$$R(x, y) = R_x(x) + R_y(y),$$

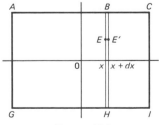

FIGURE 12.2

where $R_x(x) = T_x(\frac{1}{2}M) - T_x(x)$, $R_y(y) = T_y(\frac{1}{2}N) - T_y(y)$, which represent the portions of differential rent due to the horizontal and vertical advantages, respectively, of the point (x, y) over the corner point $(\frac{1}{2}M, \frac{1}{2}N)$ in the non-negative quadrant. Furthermore, the rent gradient is given by

$$R_x'(x) = -T_x'(x), \qquad R_y'(y) = -T_y'(y). \tag{8}$$

Again following Solow and Vickrey (1971), we shall specify the transport cost function as

$$q(z) = bz^k, \qquad b > 0, \quad k > 0. \tag{9}$$

Then the competitive rent profile can be derived explicitly as follows. First, define by $\phi(x, y)$ the cost of moving all horizontal and vertical traffic infinitesimally at (x, y), i.e.,

$$\phi(x, y) \equiv v_x(x, y)q[\delta_x(x, y)] + v_y(x, y)q[\delta_y(x, y)]. \tag{10}$$

In our special class of cases, we can write

$$\phi(x, y) = \phi_x(x) + \phi_y(y),$$

where

$$\phi_x(x) = v_x(x)b[\delta_x(x)]^k, \qquad \phi_y(y) = v_y(y)b[\delta_y(y)]^k. \tag{11}$$

In view of Eq. (3), differentiation of $\phi_x(x)$ with respect to x gives

$$\phi_x'(x) = -4(k + 1)b[\delta_x(x)]^k g(mn)^2 Nx.$$

Comparing this result with Eq. (7) and also using Eq. (8), we find

$$R_x'(x) = \phi_x'(x)/[2(k + 1)],$$

and similarly

$$R_y'(y) = \phi_y'(y)/[2(k + 1)],$$

leading to

$$R_x(x) = \phi_x(x)/[2(k + 1)], \qquad R_y(y) = \phi_y(y)/[2(k + 1)],$$

because of the boundary condition that

$$R_x(\tfrac{1}{2}M) = \phi_x(\tfrac{1}{2}M) = R_y(\tfrac{1}{2}N) = \phi_y(\tfrac{1}{2}N) = 0.$$

Thus,

$$R(x, y) = R_x(x) + R_y(y)$$
$$= [\phi_x(x) + \phi_y(y)]/[2(k + 1)] = \phi(x, y)/[2(k + 1)], \tag{12}$$

which reveals a direct relationship between $R(x, y)$ and $\phi(x, y)$ and, together with Eqs. (1), (3), (4), and (11), gives the competitive rent profile explicitly.

12.3 A Conjecture on Optimality

Consider the problem of socially optimal land use for transportation in our rectangular city model. Given the total city area K and the total business area A exogenously as $0 < A < K$, our problem is to choose the optimal values of M and N as well as the optimal functional forms of $m(x, y)$ and $n(x, y)$ from the social point of view. More specifically, the optimality problem considered here is to minimize the total transport cost incurred in the city as a whole,

$$\Phi \equiv \int_{-M/2}^{M/2} \int_{-N/2}^{N/2} \phi(x, y) \, dx \, dy,$$ (13)

with respect to M, N, $m(x, y)$, and $n(x, y)$, subject to

$$MN = K, \qquad \int_{-M/2}^{M/2} \int_{-N/2}^{N/2} m(x, y)n(x, y) \, dx \, dy = A,$$ (14)

where $\phi(x, y)$ is defined as in Eq. (10).

First, our symmetric treatment of horizontal and vertical movements implies that the optimal shape of the city is square, i.e., $M = N = L \equiv K^{1/2}$, and the optimal land use configurations are symmetric in the sense that

$$m(x, y) = m(-x, y) = m(x, -y) = m(-x, -y),$$

$$m(x, y) = n(y, x), \qquad \text{for all} \quad x \text{ and } y.$$

We cannot, however, proceed to solve for the optimal functional forms of m and n, because of the difficulty in finding equilibrium traffic patterns in general, as already pointed out. The best we can do here is to conjecture that there exists a case with certain values of K and A in which the optimal road width is constant and identically horizontally as well as vertically, i.e., $m(x, y) = m$ and $n(x, y) = n$ for all (x, y), just like the special case of optimal city size analyzed by Solow and Vickrey (1971) for which the optimal road is actually of constant width. In this special case, we have $m = n = a \equiv (A/K)^{1/2}$ from Eq. (14).

Focusing on our special case, we shall ask whether land-use decisions relying on the market value of land lead to overallocation of land to transportation, i.e., whether a public road-building agency whose decisions are based on cost–benefit analysis using competitive rents as a measure of the social cost of land tends to create too many roads in the city from the social point of view. Note first that at the optimum with $M = N = L$ and $m = n = a$, the saving in total transport cost from widening the horizontal road infinitesimally at (x, y) is

$$\frac{\partial \phi(x, y)}{\partial m(x, y)} = \frac{v_x(x)kb[v_x(x)/(1-m)]^{k-1}v_x(x)}{(1-m)^2} = \frac{k\phi(x)}{1-a},$$

and similarly the saving from widening the vertical road at (x, y) is

$$\partial\phi(x, y)/\partial n(x, y) = k\phi(y)/(1 - a),$$

where

$$\phi(z) = b(2g)^{k+1}(1 - a)^{-k}a^{4(k+1)}L^{k+1}[(\tfrac{1}{2}L)^2 - z^2]^{k+1}.$$

The benefit $S(x, y)$ arising from one unit of land at (x, y) is to be measured by the maximized amount of saving in total transport cost from converting a unit of land to transportation at (x, y), i.e.,

$$S(x, y) = k(1 - a)^{-1}\max[\phi(x), \phi(y)]. \tag{15}$$

This means that the public agency would widen horizontal (resp. vertical) roads only, whenever $|x| < |y|$ (resp. $|x| > |y|$), since cost reductions will be greater when roads with greater traffic volumes and higher densities are widened. In view of Eqs. (12) and (15), the benefit–cost ratio as calculated by the public agency becomes

$$\frac{S(x, y)}{R(x, y)} = \frac{2(k + 1)}{(1 - a)}\frac{\max[\phi(x), \phi(y)]}{\phi(x) + \phi(y)}, \tag{16}$$

which is greater than unity, if $k \geq \tfrac{1}{2}$. It can thus be stated that with a plausible value of k ($\geq\tfrac{1}{2}$), there is a tendency toward building too many roads everywhere in the square city at the optimum, just as in the Solow–Vickrey city.

It turns out, however, that this tendency toward excessive road-building is not particularly strong at the city center as Solow and Vickrey (1971) have concluded. In fact, such a tendency is stronger near the city boundaries than at the city center. First, in terms of the benefit–cost ratio, it is clear from Eq. (16) that $S(x, y)/R(x, y)$ attains its maximum value $2(k + 1)k(1 - a)^{-1}$ everywhere on the boundary lines, i.e., when $|x| = \tfrac{1}{2}L$ and/or $|y| = \tfrac{1}{2}L$; and it reaches its minimum value $(k + 1)k(1 - a)^{-1}$ everywhere on the two diagonal lines which go through the city center, i.e., when $|x| = |y|$. Second, in terms of the net benefit, i.e.,

$$S(x, y) - R(x, y) = \frac{k}{1 - a}\max[\phi(x), \phi(y)] - \frac{\phi(x) + \phi(y)}{2(k + 1)},$$

it reaches its maximum

$$\left[\frac{k}{1 - a} - \frac{1}{2(k + 1)}\right]\phi(\tfrac{1}{2}L)$$

at the midpoints of the boundary lines, i.e., either when $|x| = \tfrac{1}{2}L$ and $y = 0$ or when $x = 0$ and $|y| = \tfrac{1}{2}L$, and it attains its **minimum**

$$\left(\frac{k}{1 - a} - \frac{1}{k + 1}\right)\phi(\tfrac{1}{2}L).$$

Thus, we have established the following result.

Theorem 12.2 *In the two-dimensional square city with the elasticity of the transport cost function k equal to or greater than $\frac{1}{2}$, land-use decisions based on market rents tend to create too many roads everywhere in the city at the optimum, and the tendency toward excessive road-building is stronger near the city boundaries, especially at the midpoints of the boundary lines, than at the city center.*

An intuitive explanation why the benefit–cost ratio is maximized at the midpoints of the boundary lines and minimized at the city center is as follows. Under the present assumptions, both traffic volumes and densities are exactly the same over all horizontal roads, in particular at their midpoints, i.e., on the vertical axis, and therefore the benefits from widening horizontal roads will be the same along the vertical axis, whereas the competitive rent is highest at the center among the points on the vertical axis because of the greatest *vertical* advantage enjoyed by the center among the points on the vertical axis. Thus, the benefit–cost ratio reaches its minimum at the city center and attains its maximum at the boundary points along the vertical axis.

It should also be noted that if the vertical length of the city is negligibly small so that the transport cost of vertical movement can be ignored, then the vertical advantage of the city center over the other points on the vertical axis will disappear and the city center will virtually coincide with the midpoints of the boundary lines. Actually, this special case corresponds to the Solow–Vickrey one-dimensional city, and only in this case there is a tendency toward excessive road-building especially near the city center which is also the midpoint of the city boundary line.

12.4 Excessive City Size

So far in this chapter we have assumed that the size of the city is exogenously given. We shall now consider how the size of the city will be determined. First, we find the optimal city size which maximizes "net social product," i.e., total output minus total transport cost. Then we compare it with the city size which will be determined under perfect competition with free entry of firms.

To obtain meaningful results, let us introduce some externalities, namely, agglomeration economies in production. Specifically, we assume that the output–land ratio, defined as f, is an increasing function of city size L^2,

$$f = \beta(L^2)^\gamma, \qquad \beta > 0, \quad \gamma > 0. \tag{17}$$

Since the total amount of output for the city as a whole is fa^2L^2, the net social product will be

$$Q = f\alpha L^2 - \Phi(L) = \alpha\beta(L^2)^{\gamma+1} - \Phi(L), \tag{18}$$

where $\alpha \equiv a^2$, and Φ is defined as in Eq. (13). i.e.,

$$\Phi(L) = 2L \int_{-L/2}^{L/2} 2g\alpha^2 L[(\tfrac{1}{2}L)^2 - x^2]q_x(\quad)\,dx$$

$$= 8g\alpha^2 L^2 \int_0^{L/2} [(\tfrac{1}{2}L)^2 - x^2]q_x(\quad)\,dx \tag{19}$$

or in view of Eq. (9)

$$\Phi(L) = 8g\alpha^2 L^2 \int_0^{L/2} \left[\left(\frac{L}{2}\right)^2 - x^2\right] b \left[2g\alpha^2 L \frac{(\tfrac{1}{2}L)^2 - x^2}{1 - \sqrt{\alpha}}\right]^k dx$$

$$= B\alpha^{2(k+1)}(1 - \sqrt{\alpha})^{-k} L^{2(k+2)} L^{k+1} = CL^{3k+5}, \tag{20}$$

since we have from Eqs. (3)–(5), (10), and (13) that

$$\Phi = 2b(2g)^{k+1}(mn)^{2(k+1)}\left\{\frac{N^{k+2}}{(1-m)^k}\int_0^{M/2}[(\tfrac{1}{2}M)^2 - x^2]^{k+1}\,dx\right.$$

$$\left. + \frac{M^{k+2}}{(1-n)^k}\int_0^{N/2}[(\tfrac{1}{2}N)^2 - y^2]^{k+1}\,dy\right\}$$

$$= 2b(2g)^{k+1}(mn)^{2(k+1)}D[N^{k+2}(1-m)^{-k}(\tfrac{1}{2}M)^{2k+3}$$

$$+ M^{k+2}(1-n)^{-k}(\tfrac{1}{2}N)^{2k+3}]$$

$$= E(mn)^{2(k+1)}(MN)^{k+2}[M^{k+1}(1-m)^{-k} + N^{k+1}(1-n)^{-k}], \tag{21}$$

where B, C, D, and E are some positive constants.

From Eq. (20), the objective function becomes

$$Q = \alpha\beta L^{2(\gamma+1)} - CL^{3k+5}. \tag{22}$$

Thus, the first-order condition for maximization is[3]

$$dQ/dL = 2(\gamma+1)\alpha\beta L^{2\gamma+1} - (3k+5)CL^{3k+4} = 0, \tag{23}$$

which yields the optimal value of L as

$$L^* = [2(\gamma+1)\alpha\beta/\{(3k+5)C\}]^{1/(3k-2\gamma+3)}. \tag{24}$$

---

[3] Here we assume that the ratio of total business area to total city area, a (therefore α), is kept constant as the city expands.

The second-order condition is

$$d^2Q/dL^2 = 2(\gamma + 1)(2\gamma + 1)\alpha\beta L^{2\gamma} - (3k + 5)(3k + 4)CL^{3k+3} < 0,$$

which, together with condition (23), implies

$$3k - 2\gamma + 3 > 0. \tag{25}$$

In view of Eqs. (24) and (25), we can maintain that given $L^* > 1$, the optimal city size is increasing in the degree of agglomeration economies γ and decreasing in the degree of traffic congestion k.

Next, we examine the competitive determination of city size by assuming that the city is allowed to expand in such a way that a given proportion of business area to total city area is always maintained. Of course, we must keep the constancy of road width as well as the square shape of the city as the city expands. With no opportunity cost of land, "private transport cost" ψ_p borne by marginal producers who enter the city by expanding the city boundaries infinitesimally in all directions is equal to

$$\psi_p = \tfrac{1}{2}\Phi'(L)\big|_{\bar{q}}, \tag{26}$$

where, in view of Eq. (19),

$$\Phi'(L)\big|_{\bar{q}} = \frac{d}{dL}\left\{8g\alpha^2 L^2 \int_0^{L/2} [(\tfrac{1}{2}L)^2 - x^2]\bar{q}_x\, dx\right\}, \tag{27}$$

and

$$\Phi'(L) = \Phi'(L)\big|_{\bar{q}} + 8g\alpha^2 L^2 \int_0^{L/2} [(\tfrac{1}{2}L)^2 - x^2]\frac{dq_x(\ )}{dL}\, dx. \tag{28}$$

Since $dq_x(\ )/dL > 0$, we have

$$\Phi'(L) > \Phi'(L)\big|_{\bar{q}}. \tag{29}$$

From Eqs. (26)–(29), we see why private transport cost borne by marginal producers differs from marginal social transport cost $\Phi'(L)$. First, new producers will pay only half the transport cost generated by their entry to the city, as indicated by the factor $\tfrac{1}{2}$ in Eq. (26), because of the assumption that all transport costs are absorbed at points of traffic origin. Second, the presence of congestion costs leads to a divergence between private and social costs, as new producers will ignore the additional transport cost which they impose on others in the form of higher levels of traffic congestion. This difference can be seen from Eqs. (27)–(29).

On the other hand, marginal producers' output will be equal to a marginal increase in total output with the output–land ratio kept constant,

$$d(\bar{f}\alpha L^2)/dL = 2f\alpha L, \tag{30}$$

because marginal producers do not take account of their contribution to higher land productivities for other producers due to agglomeration economies. Under perfect competition with free entry, new producers will enter the city up to the point at which they earn zero profit,

$$2f\alpha L - \psi_p = 0, \tag{31}$$

which together with Eqs. (17) and (26) leads to

$$4\alpha\beta L^{2\gamma+1} - \Phi'(L)|_{\bar{q}} = 0. \tag{32}$$

The "competitive" city size L^c is determined by solving Eq. (32) for L.

We are now in a position to compare the competitive city size L^c with the optimal city size L^*. Noting that the competitive condition (32) becomes

$$4\alpha\beta = \Phi'(L^c)|_{\bar{q}}/(L^c)^{2\gamma+1}, \tag{33}$$

and the optimality condition (23) is rewritten as

$$2(\gamma + 1)\alpha\beta = \Phi'(L^*)/(L^*)^{2\gamma+1}, \tag{34}$$

we can find from relation (29) that

$$L^* < L^c \tag{35}$$

if

$$\gamma \leq 1, \tag{36}$$

as can be seen in Fig. 12.3. In fact, condition (36) ensures the second-order condition (25) for optimality with any value of $k \geq 0$. Thus, we have proved the following theorem.[4]

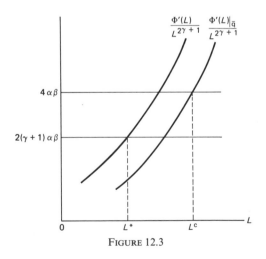

FIGURE 12.3

---

[4] For other types of optimal city size argument, see Arnott (1979) and Henderson (1977).

Theorem 12.3 *For the square city with agglomeration economies and traffic congestion, an optimal city size is uniquely determined, and free entry of competitive firms yields a city size which is greater than optimal, provided that the degree of agglomeration economies, measured by the elasticity of land productivity with respect to city size, does not exceed unity.*

REFERENCES

Arnott, R. (1979). Optimal city size in a spatial economy. *Journal of Urban Economics* **6**, 65–89.

Henderson, J. V. (1977). *Economic Theory and the Cities.* Academic Press, New York.

Kanemoto, Y. (1975). Congestion and cost-benefit analysis in cities. *Journal of Urban Economics* **2**, 246–264.

Kanemoto, Y. (1977). Cost–benefit analysis and the second best land use for transportation. *Journal of Urban Economics* **4**, 483–593.

Kraus, M. (1974). Land use in a circular city. *Journal of Economic Theory* **8**, 440–457.

Miyao, T. (1978). A note on land use in a square city. *Regional Science and Urban Economics* **8**, 371–379.

Robson, A. J. (1976). Cost benefit analysis and the use of urban land for transportation. *Journal of Urban Economics* **3**, 180–191.

Solow, R. M., and Vickrey, W. S. (1971). Land use in a long narrow city. *Journal of Economic Theory* **3**, 430–447.

CHAPTER **13**

Economics of Urban Concentration

This chapter deals with one of the most important problems in urban policy and planning, i.e., the problem of optimal concentration of urban activities in a metropolitan area. It is often asked how many business (or employment) centers ought to be created in a given urban area in view of agglomeration economies and transportation costs involved in business and commuting activities. We try to answer this question based on a square city model which is similar to the one developed in Chapter 12.

We assume two kinds of traffic, i.e., business trading traffic of producers and commuting traffic of residents, in a multicentric city with the number of business centers to be determined optimally. As in the previous chapter, we find a simple equilibrium traffic pattern which is consistent with rational route-choice behavior on the part of producers and residents. In equilibrium, total transport cost incurred in business and commuting activities can be expressed as a function of the number of business centers, given a total size of business centers in the city. On the other hand, production is assumed to display increasing returns due to agglomeration economies within each business center, so that the output–land ratio can be expressed as an increasing function of the size of each business center. This means that total output is inversely related to the number of centers. Then we determine the optimal number of business centers so as to maximize "net social product," i.e., total output minus total transport cost. Then the effects of changes in various parameters on the optimal number of centers are analyzed.

In the literature on this topic, Lave (1974) has set up a multicentric city which takes account of pure distance costs of transportation, but neither congestion costs nor agglomeration economies, and has obtained the following results: (1) the higher the cost (volume) of commuting traffic, the greater the optimal number of city centers, (2) the higher the cost (volume) of business traffic, the smaller the optimal number of centers, and (3) as the cost (volume) of commuting traffic rises relative to business traffic, the optimal number of city centers tends to jump from one to infinity.[1]

In our model with congestion costs and agglomeration economies taken into account, we show that the optimal number of business centers is increasing with the volume of commuting traffic, as Lave has shown, and also with the total size of residential area, the relative duration of commuting to business hours as well as the level of unit transport cost inside business districts, whereas the number of business centers should be smaller, as the output–land ratio is greater. On the other hand, the volume of business traffic, the total size of business districts, the level of unit transport cost within residential zones, and the width of roads will have ambiguous effects on the optimal level of concentration. We can prove, however, that if the elasticity of the transport cost function is relatively high for residential zones, the optimal number of centers is increasing with the volume of business traffic, contrary to Lave's result, because in this case the effect of congestion outweighs the effect of pure distance.

In the final section, we offer some numerical results which suggest that with no agglomeration economies in production a corner solution is likely, i.e., either a monocentric city or a perfectly dispersed city will probably be optimal. This is so in the presence of congestion costs, just as Lave's result suggests in the absence of congestion. With a modest degree of agglomeration economies, however, the optimal number of business centers seems to be increasing *gradually* with the volume (and the cost) of commuting traffic relative to business traffic, in contrast to Lave's conclusion.

13.1 The Multicentric City Model

As in the preceding chapter, we assume a square city with roads laid out densely in north–south and east–west directions, namely, vertical roads and horizontal roads, respectively. The city has a number of business–employment centers, called business districts (BDs), all of which are of the same size and of the same square shape. To each BD is attached four residential zones (RZs), which are all rectangular in shape. Each RZ has either

---

[1] For a summary of Lave's results on this topic, see Richardson (1977, pp. 98–101). For a somewhat different approach to this kind of problem, see Mills (1976).

vertical or horizontal roads leading to its adjoining BD. Any two BDs which are vertically *or* horizontally adjacent are connected by vertical *or* horizontal roads, respectively, which go through the RZs between the two BDs, as seen in Fig. 13.1, where the monocentric city case ($n = 1$) and the 4-center case ($n = 2$) are illustrated.

Each BD consists of a business area and a road area. Each unit of business area generates a certain amount of business traffic to be moved to all other units of business area in all the BDs of the city. The unit volume of business traffic is a given constant g, which is the volume of traffic to be moved from a unit of business area to another unit of business area. On the other hand, each RZ consists of a housing area and a road area. Each unit of housing area accommodates a given number of people and generates a certain amount of commuting traffic, e which has destinations uniformly distributed over all units of business area in their *nearest* BD. As for the road area, all roads have the same width $(1 - u)$ everywhere in the city, where u is a constant and $0 < u < 1$.

In order to make the number of centers variable, the city is partitioned in such a way that the same pattern of configurations is reproduced in each and every partition of the city, where the total size of the city $(M + 2L) \times (M + 2L)$, the total size of the BDs $M \times M$, the total size of the RZs $4 \times M \times L$, and the total size of the remaining (unused) area $4 \times L \times L$ are all positive constants and remain unchanged regardless of the number of partitions. Also constant are the total size of the business area $u^2 M^2$ and the total size of the road area inside the BDs $(1 - u^2)M^2$, as well as the total size of the housing area $4uML$ and the total size of the road area inside the RZs $4(1 - u)ML$. Then it is reasonable to suppose that the total cost of construction and maintenance for the city as a whole depends only on M, L, and u, and not on the number of partitions of the city. Such cost, therefore, may be safely ignored in our argument of determining the optimal level of business concentration.

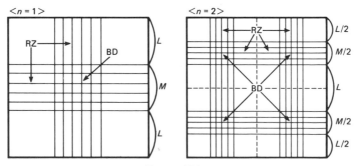

FIGURE 13.1

The basic behavioral assumption of the model is that each producer (shipper) and each commuter will choose his route in shipping or commuting so as to minimize the total transport cost of his trip in view of the existing pattern of traffic densities in the entire city. Here, the transport cost incurred per unit volume of traffic along a horizontal or vertical road at a point (x, y) is assumed to depend solely on the traffic density along the horizontal or vertical road, respectively, at the point (x, y) where x and y are the horizontal and vertical coordinates of the point in question. More specifically, denoting unit transport costs along a horizontal road and along a vertical road at (x, y) by $q_x(x, y)$ and $q_y(x, y)$, respectively, we assume that

$$q_x(x, y) = b[\delta_x(x, y)]^k, \qquad q_y(x, y) = b[\delta_y(x, y)]^k \qquad (1)$$

inside the BDs, and

$$q_x(x, y) = a[\delta_x(x, y)]^m, \qquad q_y(x, y) = a[\delta_y(x, y)]^m \qquad (2)$$

inside the RZs, where $\delta_x(x, y)$ and $\delta_y(x, y)$ are traffic densities along the horizontal and vertical roads, respectively, at (x, y), and b, k, a, and m are all positive constants. Implicit in the formulation of (1) and (2) is the assumption that the transport cost incurred by traffic along a road is not affected by any change in the density of its cross traffic. Within the model, this assumption is justifiable inside the RZs, since there is no cross traffic in the RZs by assumption. Inside the BDs, on the other hand, it may be considered as an approximation to traffic conditions in business districts with a signal system which is insensitive to the densities of traffic at cross roads.

Furthermore, we suppose that commuting hours are completely separated from business hours in such a way that commuting traffic and business traffic do not overlap anytime on any road in the city. While this assumption is not essential to obtain equilibrium traffic patterns, it will greatly simplify our analysis and computation.

13.2 Equilibrium Traffic Patterns

The first task with our two-dimensional city is to find an equilibrium traffic pattern which is consistent with individual route choice behavior. In an equilibrium, individuals must be choosing cost-minimizing paths for their trips under the present traffic pattern which, in turn, is obtained by aggregating individual route-choice behavior in the city.

As far as commuting traffic is concerned, we can concentrate on the monocentric city case ($n = 1$) without loss of generality. as the same traffic pattern is reproduced, only in a smaller scale, in each partition of the city as the number of partitions increases. Note that every commuter goes to his nearest BD only, regardless of the number of BDs in the city. We can show the following.

Theorem 13.1   *The density pattern of commuting traffic*

$$q_x^c(x, y) = q_x^c(x), \qquad q_y^c(x, y) = q_y^c(y), \qquad \text{for all } (x, y) \qquad (3)$$

is an equilibrium traffic pattern such that between any pair of origin (in a RZ) and destination (in the BD) there exists a unique "minimum-corner" path which is cost-minimizing and that the density pattern (3) is reproduced everywhere in the city if such minimum-corner paths are exclusively used by all commuters.

Proof   First, note that there is no route choice open in any RZ by assumption. As in the previous chapter, we can easily show that given the density pattern (3), all minimum-distance paths between any two points are cost-minimizing, and therefore the minimum-corner path (which is unique as seen in Fig. 13.2) between a point in a RZ and a point in the BD is cost-minimizing. Now, place the origin at the center of the BD, and consider a point (x, y) in Fig. 13.2. It is easy to see that the volume of commuting traffic passing through the point (x, y) vertically from north to south is

$$euL[(\tfrac{1}{2}M + y)M/M^2] + 2euL(\tfrac{1}{2}M - y)[(\tfrac{1}{2}M + y)/M^2], \qquad (4)$$

where e is the volume of commuting traffic generated by *each* unit of housing

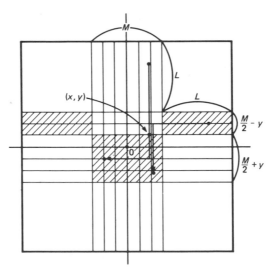

FIGURE 13.2

area. Similarly, the volume of commuting traffic passing through the point (x, y) vertically from south to north is

$$euL[(\tfrac{1}{2}M - y)M/M^2] + 2euL(\tfrac{1}{2}M + y)[(\tfrac{1}{2}M - y)/M^2]. \qquad (5)$$

Thus, the total volume of "vertical" commuting traffic at (x, y) is obtained by adding (4) and (5):

$$v_y^c(x, y) = euL\{1 + 4[(\tfrac{1}{2})^2 - (y/M)^2]\} = v_y^c(y) \qquad \text{for} \quad 0 \leq y \leq \tfrac{1}{2}M. \quad (6)$$

A symmetric argument gives the total volume of "horizontal" commuting traffic at (x, y) as

$$v_x^c(x, y) = euL\{1 + 4[(\tfrac{1}{2})^2 - (x/M)^2]\} = v_x^c(x) \qquad \text{for} \quad 0 \leq x \leq \tfrac{1}{2}M. \quad (7)$$

Therefore, the density pattern inside the BD becomes

$$\delta_x^c(x, y) = v_x^c(x, y)/(1 - u) = v_x^c(x)/(1 - u) = \delta_v^c(x)$$
$$\text{for} \quad 0 \leq x \leq \tfrac{1}{2}M, \qquad (8)$$

$$\delta_y^c(x, y) = v_y^c(x, y)/(1 - u) = v_y^c(y)/(1 - u) = \delta_y^c(y)$$
$$\text{for} \quad 0 \leq y \leq \tfrac{1}{2}M. \qquad (9)$$

As for commuting traffic inside the RZs, we place the origin at a corner of the city as shown in Fig. 13.3, and find the total volume and the density of horizontal traffic at (x, y) in a RZ as

$$v_x^c(x) = eux, \qquad \delta_x^c = eux/(1 - u) \qquad \text{for} \quad 0 \leq x \leq L. \quad (10)$$

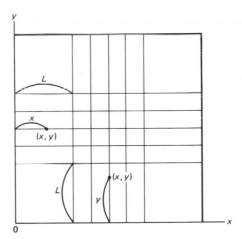

FIGURE 13.3

Similarly, for vertical traffic, we have

$$v_y^c(y) = euy, \qquad \delta_y^c(y) = euy/(1 - u) \qquad \text{for} \quad 0 \leq y \leq L. \qquad (11)$$

Symmetric results can be obtained for the remaining part of the BD and for the other RZs. Thus, the density pattern (3) is reproduced everywhere in the city.   Q.E.D.

Next, we turn to business traffic. In Chapter 12, it is shown that, denoting the volumes of horizontal and vertical business traffic at (x, y) by $v_x^s(x, y)$ and $v_y^s(x, y)$ and their densities by $\delta_x^s(x, y)$ and $\delta_y^s(x, y)$, respectively, the density pattern

$$\delta_x^s(x, y) = \delta_x^s(x), \qquad \delta_y^s(x, y) = \delta_y^s(y) \qquad \text{for all} \quad (x, y) \qquad (12)$$

is an equilibrium traffic pattern in the monocentric city case ($n = 1$), where

$$v_x^s(x) = 2gu^4M[(\tfrac{1}{2}M)^2 - x^2] \qquad \text{for} \quad 0 \leq x \leq \tfrac{1}{2}M,$$
$$v_y^s(y) = 2gu^4M[(\tfrac{1}{2}M)^2 - y^2] \qquad \text{for} \quad 0 \leq y \leq \tfrac{1}{2}M, \qquad (13)$$

and

$$\delta_x^s(x) = v_x^s(x)/(1 - u) \qquad \text{for} \quad 0 \leq x \leq \tfrac{1}{2}M,$$
$$\delta_y^s(y) = v_y^s(y)/(1 - u) \qquad \text{for} \quad 0 \leq y \leq \tfrac{1}{2}M. \qquad (14)$$

In this case there is no business traffic outside the BD. It turns out that the main result for $n = 1$ will survive when $n > 1$. In fact, we can prove the following.

Theorem 13.2   *The density pattern of business traffic*

$$\delta_x^s(x, y) = \delta_x^s(x), \qquad \delta_y^s(x, y) = \delta_y^s(y) \qquad \text{for all} \quad (x, y) \qquad (15)$$

is an equilibrium traffic pattern for any $n \geq 1$ such that if, among the cost-minimizing paths between any two points in any BD(s), the minimum corner paths are exclusively and equally utilized, then the density pattern (15) is reproduced everywhere in the city.

Proof   As shown in the previous chapter, it is easy to see that, given the density pattern (15), the cost-minimizing paths between any two points in any BD(s) coincide with the minimum-distance paths between the two points, not only for $n = 1$, but also for $n > 1$, because the transport cost incurred along a horizontal (or vertical, resp.) road depends only on the horizontal (or vertical, resp.) coordinate of the road everywhere in the city. Then, the minimum-corner paths which are the two outer paths along the boundaries of the smallest rectangle including traffic origin and destination at its corners

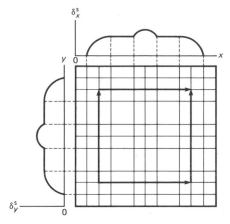

FIGURE 13.4

are obviously distance-minimizing and therefore cost-minimizing, as seen in Fig. 13.4. By ignoring the portions of the roads in the RZs, we form a "consolidated" BD as illustrated in Fig. 13.5, and find that the main result for $n = 1$ remains valid for the consolidated BD. Thus, we still have (13) and (14) inside the BDs for any $n > 1$ with suitable adjustment in reading the horizontal and vertical coordinates of each point in the BDs. As for the RZs, it is clear from Fig. 13.4 that since all roads have the same constant width, the traffic density in the segment of a horizontal road between any two adjacent BDs is constant and is equal to the density of horizontal traffic at a western (or eastern, resp.) boundary point of the BD lying east (or west, resp.) of the road segment considered. Thus, traffic densities along a horizontal road are independent of the vertical coordinate of the road. A symmetric argument applies to vertical roads, and the density pattern (15) is reproduced.   Q.E.D.

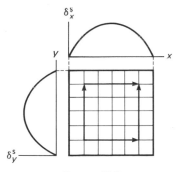

FIGURE 13.5

13.3   Total Transport Cost

Let $C_b^c(n)$ denote the total commuting cost incurred inside the BDs for n. In the case that $n = 1$, we find from Eqs. (6)–(9) that

$$
\begin{aligned}
C_b^c(1) &= 4 \int_0^{M/2} \int_0^{M/2} \{v_x^c(x)b[\delta_x^c(x)]^k + v_y^c(y)b[\delta_y^c(y)]^k\} \, dx \, dy \\
&= \frac{4bM(euL)^{k+1}}{(1-u)^k} \int_0^{M/2} \left\{ 1 + 4\left[\left(\frac{1}{2}\right)^2 - \left(\frac{x}{M}\right)^2 \right] \right\}^{k+1} dx \\
&= \frac{4b(euL)^{k+1}}{(1-u)^k} M^2 \alpha(k),
\end{aligned}
\tag{16}
$$

where $\alpha(k)$ is an increasing function of k. In the general case that $n \geq 1$, the number of BDs is n^2, and thus

$$
C_b^c(n) = n^2 [4b(eu)^{k+1}/(1-u)^k](L/n)^{k+1}(M/n)^2\alpha(k) = (1/n)^{k+1} C_b^c(1). \tag{17}
$$

Let $C_r^c(n)$ denote the total commuting cost incurred inside the RZs for n. In the case that $n = 1$, it follows from Eqs. (10) and (11) that

$$
C_r^c(1) = 4M \int_0^L v_x^c(x)a[\delta_x^c(x)]^m \, dx = \frac{4aM(eu)^{m+1}}{(1-u)^m} \frac{L^{m+1}}{m+2}, \tag{18}
$$

and in the general case

$$
C_r^c(n) = n^2 \frac{4a(eu)^{m+1}}{(1-u)^m} \left(\frac{M}{n}\right)\left(\frac{L}{n}\right)^{m+2} \frac{1}{m+2} = \left(\frac{1}{n}\right)^{m+1} C_r^c(1). \tag{19}
$$

Turning to business traffic, let $C_b^s(n)$ be the total shipping cost incurred inside the BDs for n. As seen in the proof of Theorem 13.2, $C_b^s(n)$ remains constant for all $n \geq 1$, so that from Eqs. (13) and (14)

$$
\begin{aligned}
C_b^s(n) = C_b^s(1) &= 4 \int_0^{M/2} \int_0^{M/2} \{v_x^s(x)b[\delta_x^s(x)]^k + v_y^s(y)b[\delta_y^s(y)]^k\} \, dx \, dy \\
&= \frac{4b(2gu^4)^{k+1}}{(1-u)^k} M^{k+2} \int_0^{M/2} \left[\left(\frac{M}{2}\right)^2 - x^2 \right]^{k+1} dx \\
&= \frac{4b(2gu^4)^{k+1}}{(1-u)^k} M^{k+2} \left(\frac{M}{2}\right)^{2k+3} \beta(k) = \frac{b(gu^4)^{k+1}}{(1-u)^k} M^{3k+5} \left(\frac{1}{2}\right)^k \beta(k),
\end{aligned}
\tag{20}
$$

where $\beta(k)$ is a decreasing function of k.

The total shipping cost incurred inside the RZs for n, denoted by $C_r^s(n)$, can be obtained as follows. In the case $n = 1$, it is obvious that

$$
C_r^s(1) = 0. \tag{21}
$$

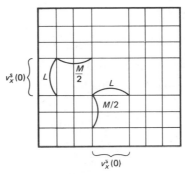

FIGURE 13.6

In the case $n = 2$, we have $v_x^s(0) = 2gu^4M(\tfrac{1}{2}M)^2$, and thus from Fig. 13.6,

$$C_r^s(2) = 4(\tfrac{1}{2}M)Lv_x^s(0)a\big[\delta_x^s(0)\big]^m = \big[2MLa/(1-u)^m\big](\tfrac{1}{2})^{m+1}(gu^4M^3)^{m+1} = (\tfrac{1}{2})^m A, \tag{22}$$

where

$$A \equiv \big[a(gu^4)^{m+1}/(1-u)^m\big]LM^{3m+4}. \tag{23}$$

In the case $n = 3$, we obtain

$$v_x^\varepsilon(\tfrac{1}{6}M) = 2gu^4M\big[(\tfrac{1}{2}M)^2 - (\tfrac{1}{6}M)^2\big],$$

and find from Fig. 13.7,

$$\begin{aligned}C_r^s(3) &= 12(\tfrac{1}{3}V)(\tfrac{2}{3}L)v_x^s(\tfrac{1}{6}M)a\big[\delta_x^s(\tfrac{1}{6}M)\big]^m \\ &= \lfloor 8MLa/3(1-u)^m\rfloor\{2gu^4M\big[(\tfrac{1}{2}M)^2 - (\tfrac{1}{6}M)^2\big]\}^{m+1} = \tfrac{8}{3}(\tfrac{4}{9})^{m+1}A. \quad (24)\end{aligned}$$

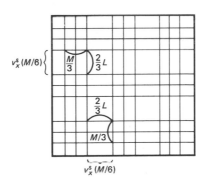

FIGURE 13.7

Similarly, we can calculate

$$C_r^s(4) = \left[(\tfrac{1}{2})^{m+1} + 2(\tfrac{3}{8})^{m+1}\right]A, \tag{25}$$

$$C_r^s(5) = \tfrac{8}{5}\left[(\tfrac{8}{25})^{m+1} + (\tfrac{12}{25})^{m+1}\right]A, \tag{26}$$

$$C_r^s(6) = \left[\tfrac{1}{3}(\tfrac{1}{2})^m + \tfrac{4}{3}(\tfrac{4}{9})^{m+1} + \tfrac{4}{3}(\tfrac{5}{18})^{m+1}\right]A, \tag{27}$$

and in general

$$C_r^s(n) = \phi(n;m)A, \tag{28}$$

where $\phi(n;m)$ behaves as in Fig. 13.8, which indicates that ϕ is decreasing in n (≥ 2) for $m = 2$, and this seems to be the case for any $m \geq 2$. This means that if the elasticity of the transport cost function is relatively high for the RZs, a larger number of partitions will reduce the total shipping cost incurred inside the RZs, because more partitioning will lead to a lower level of congestion, which outweighs higher distance cost due to dispersion.

Finally, let $\Phi(n)$ be the total transport cost incurred in the whole city for n. Denoting the relative duration of commuting to business hours by z, we find from Eqs. (17), (19), (20), and (28) that

$$\Phi(n) = z\left[C_b^c(n) + C_r^c(n)\right] + \left[C_b^s(n) + C_r^s(n)\right]$$

$$= z\left[(1/n)^{k+1}C_b^c(1) + (1/n)^{m+1}C_r^c(1)\right] + C_b^s(1) + \phi(n;m)A. \tag{29}$$

For our later use, we take up a special case with $k = m$, in which case it follows from Eqs. (17–29) that

$$\Phi(n) = \left[M/(1-u)^k\right]\{4z(1/n)^{k+1}(euL)^{k+1}\left[bM\alpha(k) + (aL/\{k+2\})\right]$$

$$+ M^{3k+3}(gu^4)^{k+1}\left[bM(\tfrac{1}{2})^k\beta(k) + aL\phi(n;k)\right]\}. \tag{30}$$

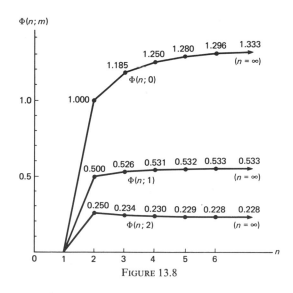

FIGURE 13.8

13.4   Net Social Product

Let us now turn from the cost side to the production side of the problem. We shall introduce agglomeration economies in production with respect to the size of each business center. Due to production externalities which are effective within each BD, but not across BDs (e.g., local public goods, personal exchange of information, etc.), production displays increasing returns to scale with respect to the total size of the business area in each BD so that we can express the amount of output produced from each unit of business area, f, as an increasing (nondecreasing) function of the size of the business area in each BD,

$$f = f[u^2(M/n)^2], \qquad f'(\ ) \geqq 0. \tag{31}$$

Specifically, we assume

$$f = s[u^2(M/n)^2]^\gamma, \tag{32}$$

where s and γ are constants and $s > 0$ and $\gamma \geq 0$. The coefficient γ may be called the degree of agglomeration economies.

Then, the total output produced in the whole city is equal to

$$Y(n) = n^2 s[u^2(M/n)^2]^\gamma u^2(M/n)^2 = s(uM)^{2\gamma+2}(1/n)^{2\gamma}. \tag{33}$$

Define "net social product" as total output minus total transport cost for the city as a whole,

$$Q(n) \equiv Y(n) - \Phi(n). \tag{34}$$

Thus, our problem is to choose the optimal value of n so as to maximize

$$Q(n) = s(uM)^{2\gamma+2}(1/n)^{2\gamma} - \{z[(1/n)^{k+1}C_b^c(1)$$
$$+ (1/n)^{m+1}C_r^c(1)] + C_b^s(1) + \phi(n;m)A\}. \tag{35}$$

In order to examine the effects of changes in various parameters on the optimal value of n, we suppose that there exists an interior optimum n^* such that $1 < n^* < \infty$, and we approximate such an optimum position by treating n as continuous rather than discrete in Eq. (35). Then at the optimum n^*, the first-order condition

$$Q'(n) = -2\gamma s(uM)^{2\gamma+2}(1/n)^{2\gamma+1} + z[(k+1)(1/n)^{k+2}C_b^c(1)$$
$$+ (m+1)(1/n)^{m+2}C_r^c(1)] - \phi_n(n;m)A = 0 \tag{36}$$

and the second-order condition

$$Q''(n) < 0 \tag{37}$$

must be met, where $\phi_n(\ )$ is the partial derivative of ϕ with respect to n. Now we prove the following proposition.

Theorem 13.3   *The optimal value of n is increasing in e, L, z, and b, and is decreasing in s; i.e., the optimal number of business centers is increasing with the volume of commuting traffic, the total size of the RZs, the relative duration of commuting to business hours, and the level of unit transport cost inside the BDs, whereas the number of business centers should be smaller as the output-land ratio is higher.*

Proof   Rewriting condition (36) as $Q_n(n; w) = 0$ with any parameter w, we find $dn/dw = -Q_{nw}(\ )/Q_{nn}(\ )$, where subscripts indicate partial differentiation. Since $Q_{nn} < 0$ from (37), dn/dw has the same sign as Q_{nw}. Thus, it follows from (36) that

$$Q_{ne} = z(k + 1)(1/n)^{k+2}\, \partial C_b^c(1)/\partial e + z(m + 1)(1/n)^{m+2}\, \partial C_r^c(1)/\partial e > 0, \quad (38)$$

in view of the fact that both $C_b^c(1)$ and $C_r^c(1)$ are increasing in e from (16) and (18). We also have

$$Q_{nL}L = z(k + 1)(1/n)^{k+2}(k + 1)C_b^c(1) + z(m + 1)(1/n)^{m+2}(m + 2)C_r^c(1)$$
$$- \phi_n(\ )A > 0 \qquad \text{from (36)}, \quad (39)$$

$$Q_{nz} = (k + 1)(1/n)^{k+2}C_b^c(1) + (m + 1)(1/n)^{m+2}C_r^c(1) > 0, \quad (40)$$

$$Q_{nb} = z(k + 1)(1/n)^{k+2}\, \partial C_b^c(1)\partial b > 0 \qquad \text{from (16)}, \quad (41)$$

$$Q_{ns} = -2\gamma(uM)^{2\gamma+2}(1/n)^{2\gamma+1} < 0. \quad \text{Q.E.D.} \quad (42)$$

On the other hand, the effects of changes in the volume of business traffic, the total size of the BDs, the level of unit transport cost inside the RZs, and the width of roads on the optimal value of n are all ambiguous in general, because these parameters affect the total shipping cost incurred in the RZs, which may be increasing or decreasing in n, depending on the value of m. However, we can prove the following.

Theorem 13.4   *Under the present assumptions, we have* (i) $dn^*/dg \gtreqless 0$, *as* $\phi_n(n^*; m) \gtreqless 0$, (ii) $dn^*/dM < 0$, *if* $\phi_n(n^*; m) \geqq 0$, (iii) $dn^*/da > 0$, *if* $\phi_n(n^*; m) \leqq 0$, *and* (iv) $dn^*/du > 0$, *if* $\phi_n(n^*; m) \leqq 0$ *and* $1 + k \geqq 1 + m \geqq 2(1 + \gamma)$.

Proof   (i)   The result follows from the fact that $Q_{ng} = -\phi_n(n; m)\, \partial A/\partial g$, together with $\partial A/\partial g > 0$ from (23).

(ii) $Q_{nM}(\tfrac{1}{2}M) = -(1 + \gamma)2\gamma s(uM)^{2\gamma+2}(1/n)^{2\gamma+1} + z(k + 1)(1/n)^{k+2}C_b^c(1) + z(m + 1)(1/n)^{m+2}\tfrac{1}{2}C_r^c(1) - \phi_n(n; m)[\tfrac{1}{2}(3m + 4)]A < 0$, if $\phi_n(n; m) \geqq 0$, in view of (36).

(iii) $Q_{na} = z(m + 1)(1/n)^{m+2}\, \partial C_r^c(1)/\partial a - \phi_n(n; m)\, \partial A/\partial a > 0$, if $\phi_n(n; m) \leqq 0$, since both $C_r^c(1)$ and A are increasing in a from (18) and (23).

(iv) $Q_{nu}u/[2(1 + \gamma)] = -2\gamma s(uM)^{2\gamma+2}(1/n)^{2\gamma+1} + z(k + 1)(1/n)^{k+2}DC_b^c(1) + z(m + 1)(1/n)^{m+2}EC_r^c(1) - \phi_n(n; m)\{E + [3(1 + m)/2(1 + \gamma)]\}A > 0$, if $\phi_n(\ ) \leqq 0$

and $1 + k \geq 1 + m \geq 2(1 + \gamma)$, since $D \equiv \frac{1}{2}\{1 + [k/(1 - u)]\}/(1 + \gamma) > 1$ and $E \equiv \frac{1}{2}\{1 + [m/(1 - u)]\}/(1 + \gamma) > 1$, and thus $Q_{nu} > 0$ in view of (36). Q.E.D.

According to our calculation as shown in Fig. 13.8, $\phi(n; m)$ seems to be increasing in n for $m \leq 1$, and decreasing in n (≥ 2) for $m \geq 2$. Then we may interpret Theorem 13.4 as follows.

Theorem 13.5 *The optimal number of business centers is smaller as the volume of business traffic is greater and as the total size of the BDs is larger, provided that the elasticity of the transport cost function is relatively low for the RZs ($m \leq 1$). If, however, the elasticity is relatively high ($m \geq 2$), the optimal number of centers is increasing with the volume of business traffic and also with the level of unit transport cost in the RZs. Furthermore, the narrower the roads, the greater the optimal number of centers in the case that the elasticity of the transport cost function is relatively high everywhere and is higher in the BDs than in the RZs ($k \geq m \geq 2$) and that the degree of agglomeration economies is relatively low ($\gamma \leq \frac{1}{2}$).*

13.5 Numerical Examples

Let us provide some numerical examples in order to reexamine Lave's conclusion that "when the cost of commuting rose enough in conjunction with a fall in freight transportation cost to induce a movement away from a single city center, one immediately noticed a jump to a vast number of centers, rather than a gradual increase to two, and then three, etc." (see Lave, 1974, p. 57). Noting that in our model with congestion, the unit cost of commuting is increasing with the volume of commuting traffic, we define a parameter

$$\theta = \tau e^2, \tag{43}$$

which represents the "normalized" volume of commuting traffic relative to business traffic, where the latter volume along with other parameter values is given as

$$g = 0.01, \quad k = m = b = 1, \quad a = 0.5,$$
$$M = 20, \quad L = 30, \quad u = 0.5, \quad s = 300. \tag{44}$$

Having $k = m$ in this special case, we can make use of Eq. (30) in calculating $\Phi(n)$, and find net social product as

$$Q(n; \theta, \gamma) = (30{,}000)(10/n)^{2\gamma} - [(1{,}212{,}000)\theta(1/n)^2$$
$$+ 4000 + (15{,}000)\phi(n; 1)], \tag{45}$$

where we consider the parameter values

$$\gamma = 0, \tfrac{1}{32}, \tfrac{1}{8}, \qquad \theta = 0.006, 0.008, 0.01, 0.05. \tag{46}$$

In the absence of agglomeration economies, i.e., $\gamma = 0$, maximization of Q is equivalent to minimization of Φ. For this case, a list of total transport costs is given in Table 13.1, where the minimum cost is indicated by the asterisk for each given θ. The result here suggests that in the case that $\gamma = 0$, a corner solution is very likely, i.e., either a monocentric city of a perfectly dispersed city will probably be optimal, just as Lave concluded in his model with no traffic congestion. In the presence of agglomeration economies ($\gamma > 0$), the optimal number of centers is determined so as to maximize net social product $Q(n; \theta)$. As seen in Tables 13.2 and 13.3, the optimal number of business centers seems to be increasing gradually with the volume (and

TABLE 13.1

TOTAL TRANSPORT COST Φ FOR $\gamma = 0$

θ	$n = 1$	$n = 2$	$n = 3$	$n = 4$	$n = 5$	$n = 6$	$n = \infty$
0.006	11,272*	13,318	12,698	12,420	12,270	12,197	12,000
0.008	13,696	13,924	12,968	12,571	12,368	12,264	12,000*
0.01	16,120	14,530	13,237	12,723	12,465	12,332	12,000*
0.05	28,240	17,560	14,583	13,480	12,950	12,668	12,000*

TABLE 13.2

NET SOCIAL PRODUCT Q FOR $\gamma = \tfrac{1}{32}$

θ	$n = 1$	$n = 2$	$n = 3$	$n = 4$	$n = 5$	$n = 6$
0.006	23,348*	19,832	19,642	19,321	19,050	18,763
0.008	20,924*	19,226	19,372	19,169	18,952	18,695
0.01	18,500	18,620	19,103*	19,017	18,855	18,628
0.05	6380	15,590	17,757	18,260	18,370*	18,292

TABLE 13.3

NET SOCIAL PRODUCT Q FOR $\gamma = \tfrac{1}{8}$

θ	$n = 1$	$n = 2$	$n = 3$	$n = 4$	$n = 5$	$n = 6$
0.006	42,068*	31,532	27,802	25,290	23,400	21,883
0.008	39,644*	30,926	27,532	25,139	23,302	21,816
0.01	37,220*	30,320	27,263	24,987	23,205	21,748
0.05	25,100	27,290*	25,917	24,230	22,720	21,412

the cost) of commuting traffic relative to business traffic, especially in the case that $\gamma = \frac{1}{8}$. This result is in contrast to Lave's conclusion in his model with no agglomeration economies in production.

In summary, we have shown the following.

Theorem 13.6 *According to Tables* 13.1–13.3, *we find that with no agglomeration economies either a monocentric city or a perfectly dispersed city will be optimal, depending on the volume of commuting traffic relative to business traffic, but with a modest degree of agglomeration economies the optimal number of business centers will gradually increase with the volume of commuting traffic relative to business traffic.*

REFERENCES

Lave, L. B. (1974). Urban externalities. In *Papers from the Urban Economic Conference 1973,* CP9, Vol. 1, pp. 39–95. Centre for Environmental Studies, London.
Mills, E. S. (1976). Planning and market processes in the urban models. In *Public and Urban Economics* (R. E. Grieson, ed.), pp. 313–329. Lexington Books, Lexington, Massachusetts.
Richardson, H. W. (1977). *The New Urban Economics: and Alternatives.* Pion Ltd., London.

Index